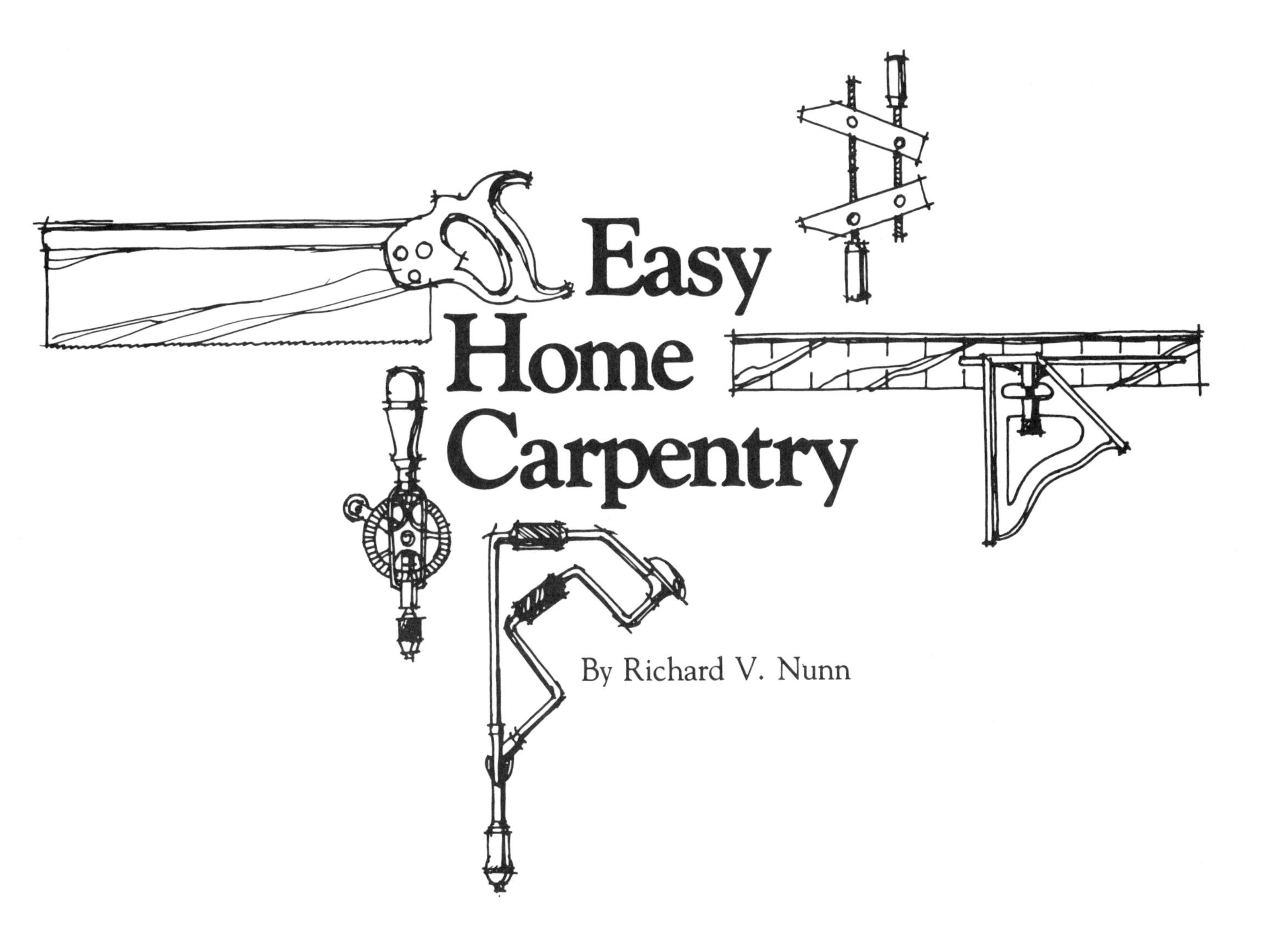

Easy Home Carpentry

By Richard V. Nunn

Book Division of The Progressive Farmer Company
P.O. Box 2463, Birmingham, Alabama 35202

ISBN: 0-8487-0373-1

Library of Congress Catalog Card Number: 74-18643

Manufactured in the United States of America

First Printing 1975
Second Printing 1976

Easy Home Carpentry

Editor: Karen Phillips
Cover Photograph: Taylor Lewis

Contents

Introduction 4

Finding a Place to Work 5
- Workshop planning 5
- Space planning 5
- Light, heat, power 6
- Low-maintenance materials 7
- Workbench and tool board 7
- Other considerations 7

Basic Hand Tools and How to Use Them 9
- Hammer how-to 10
- Saw how-to 16
- Chisel how-to 22
- Keeping your chisels sharp and in shape 25
- Plane how-to 26
- Screwdriver how-to 32
- Hole boring how-to 37
- How to select bits and drills 40
- Measuring and marking how-to . . . 42
- Pliers, wrenches, and clamps 45

Basic Fastening Techniques 51
- Common nails 51
- Finishing nails 52
- Common screws 53
- Specialty fasteners you'll use often . . 54
- Adhesives are fasteners 57
- Picking the right adhesive for the job 58
- Wood joinery techniques 58

Choosing and Using Basic Building Materials 63
- Dimension lumber and boards 63
- Plywood 63
- Hardboard 64
- Moldings 64
- Gypsum wallboard 64
- Used lumber is a bargain 65
- Tips on working with plywood 65
- Tips on working with hardboard . . . 67
- Basic lumber sizes 69

Basic Framing and Finishing Techniques 70
- Application basics for gypsum wallboard 87
- Furring strips for ceiling tile and paneling 90

Index 95

Introduction

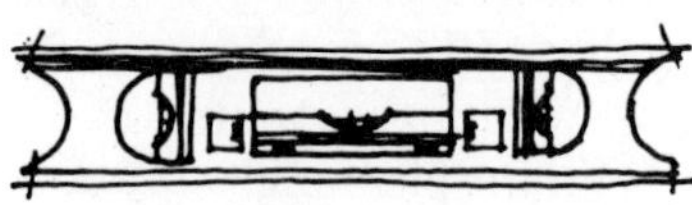

Easy Home Carpentry serves a dual purpose.

1. It is a "fun" book for the homeowner/handyman who likes to work with tools and wood, and wants to learn more of the basic skills and techniques.

2. It is a "duty" how-to book for the homeowner/handyman who would really rather be golfing, playing tennis, or resting in a hammock than pounding nails or sawing 2 by 4's for a partition wall in the basement. But, alas, the nails need to be pounded and the wood sawed, because it is cheaper these days to do the job yourself than to have a professional do it—if a professional can be found and hired.

Either way, *Easy Home Carpentry* is a basic book of carpentry. It will show you how to handle a hammer, how to saw, how to plane, how to bore holes, and how to fasten wood; you will also learn what materials to use where — from building bookshelves for your living room to finishing off a new porch.

You won't find do-it-yourself projects in this book. What we show is how to *use* tools and materials for almost any do-it-yourself project that is undertaken. Nor is *Easy Home Carpentry* a tool or material book. You won't find a special tool to drive nails when a hammer will do, or a suggestion of how to work with cherry plywood at $30 or so per sheet when regular fir plywood will fill the bill and save some money.

Whether for "fun" or "duty," the idea of *Easy Home Carpentry* is to help you do repairs, remodeling, and other home building projects more quickly and efficiently at a cost you can afford. And, as an added bonus, we think you will like the results that the proper use of tools and materials produce. All of it is easier than you might think.

Finding a Place to Work

Home maintenance and improvement is not a one-shot deal; you don't just add on a room, put up a ceiling, or panel a room. Home improvement is a continual project, although at the outset you may think differently. This is why it is important for you to assemble a working area—a workshop with a bench is essential as well as tool organization. A place to work is the first step to easy carpentry.

Your workshop does not have to be costly or beautifully appointed, but it has to be comfortable—a place in which you like to work and relax. Your workshop should also be conveniently located in relationship to the projects you'll be doing, and easy to keep clean. A wobbly old table in the corner of a room won't do for a workshop; in fact, an inadequate layout is a definite handicap.

Plan your workshop from the outset, however humble; you'll find your projects will be easier to build and will have a professional look to them. If you work on a wobbly bench in a poorly lighted room, your projects will be wobbly and lack professional luster.

Workshop planning

Somewhere in your home there is a spot for a workshop, even if you live in an apartment. Here are some ideas for shop location:

- In a basement room—perhaps in the furnace area
- Under the basement stairsteps
- In a large utility closet
- The back section of your garage
- An addition to your carport
- In the attic
- In a freestanding tool shed
- A corner of the recreation or family room

Regardless of where the workshop is located, you must organize it—even if the shop is just a large toolbox you move to an open area. The major ingredients that go into a smooth-running shop are smart space planning, light, heat, adequate power (for power tools), and low-maintenance materials. These factors apply to a large shop as well as to one that fits between the front bumper of your car and the back of the garage wall.

Space planning

Plan workshop space for flexibility and expansion. In the beginning, allow as much space as you can for the shop. Of course, the more space you have, the better off you are. At the same time, think about the future—an adjoining area where you may "borrow" floor space for fabricating or assembling large projects. Don't overlook locking casters for power tools, when you add them to your tool selection. On wheels, tools may be rolled into temporary working quarters near the main shop complex.

There is no rule of thumb on shop size. However, we would recommend a minimum of 8 by 10 feet, since you need elbow room to turn sheets of plywood and lengths of dimension lumber. You cannot, of course, squeeze this much space from an under-the-stairsteps shop. But you can use areas adjoining the stairsteps for major woodworking projects.

Assuming you have the room, plan on about 24 square feet of floor area for a workbench that stands against a wall in minimum space. You will need a tool board, but this will hang on the wall. You'll also need about 16 square feet of floor space for shelving to hold containers of nails, screws, bolts, paint, stain, and so forth.

If you have plenty of space, consider an island workbench layout. Here, you can work around all 4 sides of the bench, which is a definite advantage. Or, try a peninsula workbench that provides 3 working sides.

Your future plans may call for power tools — a table or radial arm saw, drill press, band saw, lathe, and sander. If so, lay out the shop in zones. Power saws should be located near the workbench. This enables you to take advantage of the workbench top to support large pieces of material when you are using the saw. A jointer and drill press are secondary to the saws. A lathe, sander, and grinder may be located in another zone, since these tools are not in use as much as the sawing, planing, and drilling equipment.

The secret is to zone areas so that the shop has an assembly line procedure for projects. The work flow then becomes smoother and easier; tools are related.

Light, heat, and power

For safety, you must have adequate lighting in the shop, whether or not you have power tools. We recommend a minimum of one 40-watt, 2-tube fluorescent light fixture. Fluorescent lighting produces an even lighting throughout the shop area. You can increase this light level by painting or covering the workshop walls with a light wall covering, or white paint. In zoned areas, spotlighting teamed with fluorescent lighting is recommended.

Your workshop should have adequate heat so you will not have to be restricted by a jacket or heavy sweater while you work. Heat is also important when you work with glues and finishes.

Although, at the beginning, you may not have the budget for power tools, you should provide wiring for those power tools that you probably will want to add at a later date. Plan on 2 circuits—one for power equipment, the other for lighting. We recommend a 20- to 30-amp circuit for tools running on motors of ¾ horsepower or more. A 15-amp

Minimum space workshop occupies one corner of this basement, utilizing about 80 square feet of space. The tool board is a sheet of 4 by 8 feet tempered, perforated hardboard mounted to the wall with 3 furring strips. The strips were sealed with clear, penetrating sealer before they were mounted to the wall with toggle bolts.

circuit usually is adequate for lighting. Your local utility company can help you with power planning, which really involves the number of tools you'll have in your home workshop.

Low-maintenance materials

A workshop that is easy to keep clean usually is kept clean. A clean shop is safer; it adds to your enjoyment and tends to help you produce better projects.

Since most basement walls are concrete block or reinforced concrete, paint the walls with a semigloss finish. Bare concrete tends to collect sawdust unless it has been sealed with a paint covering. If you want to be fancy, you can cover the walls with a decorative (or plain) hardboard paneling. Paint the plain hardboard a light color. The hardboard should be the tempered type to resist moisture.

Acoustical ceiling tile in a light color is easier to keep clean than open floor joints. The ceiling tile also has the advantage of dampening the sound of tools at work. When you graduate to power tools, you can also mount the motors on rubber pads to help deter the noise.

Paint or tile the workshop floor. The color should be light and the surface smooth so that the floor is easy to keep clean with only an occasional sweeping or vacuuming.

If at all possible, have a door on your workshop and keep the door locked when the shop is not being used. A door provides safety: kids love to use tools, especially powered ones. Hopefully, a locked door will keep the kids out until you are there to supervise the activity. At least, the door lock will slow the youngsters down. A door also provides some sound conditioning and prevents the bulk of the sawdust from filtering out of the shop into other rooms.

Workbench and tool board

You can buy workbenches at most building material outlets, hardware stores, and general merchandise stores with a catalog service. Or, you can build your own workbench. Either way, the workbench must be sturdy since it will take a lot of pounding and punishment. A wobbly, undersized workbench is a waste of money. Unfortunately, there are many of these on the market, so be sure to shop carefully.

The workbench pictured in this chapter was made from a 6/8 solid core door, and the cost, on sale, was $13. The legs are metal angle iron, although 4 by 4-inch wood timbers (bolted or lag-screwed together) could be used. The bench is hip high, about 42 to 44 inches, and the top of it is covered with ¼-inch tempered hardboard. The tool board is a ¼-inch tempered, perforated hardboard.

If your workbench is too low, you will have back problems from hunching over; if the bench is too high, your hands and arms will be restricted. The "hipbone high" formula generally is correct for a workbench or other shop table arrangement.

Other considerations

Even with hand tools, you will create a lot of dirt and dust in your workshop; that's why adequate ventilation within the shop is very important. Adequate ventilation is also necessary when you use paint, stains, and some glues. You can control dust and fumes with an exhaust fan that is mounted into a window (the way a clothes dryer is vented) or spliced into a stove-pipe duct.

Be aware of noise control in the workshop—your family will be aware of the noise. Sound deadening board (similar to insulation board sheathing) in back of the paneling will help control sound, as will acoustical ceiling tile, which absorbs about 60 percent of the noise that strikes it.

Tempered hardboard makes an excellent workbench top. It is simply nailed to the top surface of the bench, here a solid core door. Use 5d finishing nails to fasten the hardboard down; countersink the nails. When the hardboard becomes damaged, it may be removed, turned over, and renailed. When this side becomes worn, the material may be cut into scrap, and a new hardboard surface may be installed. The hardboard is tough and very easy to keep clean with a damp cloth.

Locking hooks for perforated hardboard to hold tools won't pull out once they are snapped into position. The hooks also may be unlocked and repositioned. A wide variety of hooks are available for hanging almost any tool.
Regular hardboard hooks have tiny wire locks that prevent them from moving when a tool is removed or replaced from the tool board.

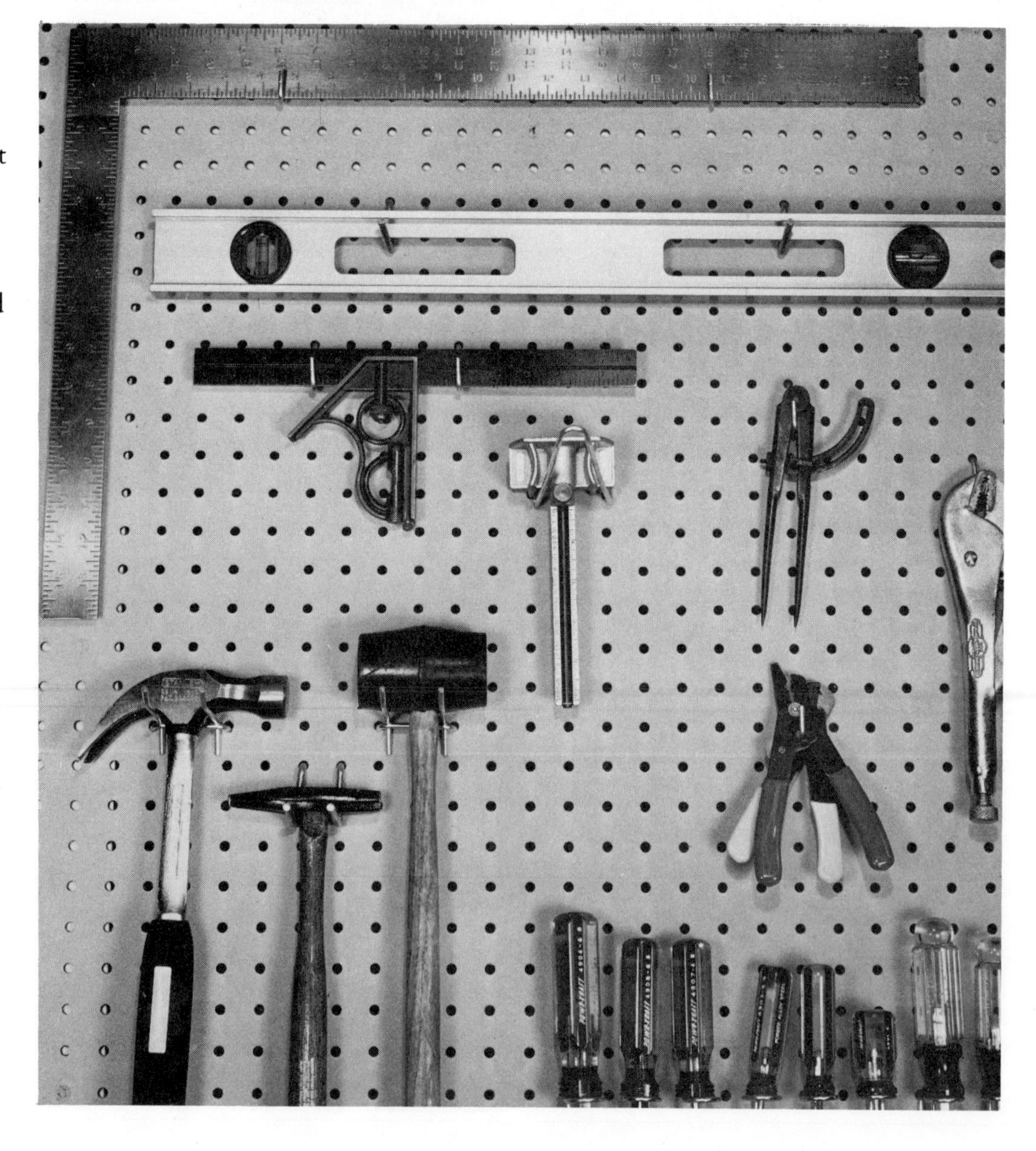

Organization is the key to workshop planning, and this even includes placement of tools over the workbench. Keep different types of tools assembled in a certain area: hammers together, planes in one place, screwdrivers side by side, and measuring and marking tools in another area. Some tools, such as levels and squares, can occupy corner sections of the tool board, utilizing otherwise wasted space.

Basic Hand Tools and How to Use Them

Unless you have a lot of money to spend, it is best to equip your workshop in stages. First, buy the basic hand tools: a hammer, saw, plane, brace and assortment of bits, pliers, and a set of screwdrivers. Then, as your budget allows, add to these basics. From here, you may graduate to the supplementary hand tools, then into power equipment.

There are just 3 power tools we recommend at the start: a ¼-inch portable electric drill, an electric sander, and a power saw. These tools, used very frequently in carpentry, home maintenance, and repair, can save plenty of time, and they are not too costly.

The basic tools recommended here are for so-called "beginners," simply because a homeowner/handyman first has to learn to work with the basics. Then, as your skills increase and you become more confident with the basics, you can add to your initial investment.

Finally, as you develop into a first-rate carpenter, you will be more selective in the purchase of tools, since you will know exactly what your tool needs will be.

From start to finish, *do not* buy or use cheap tools. Quality tools cost only a few dollars more than the cheap ones, but they will last a lifetime. Too, quality tools are easier to use, and they are safer.

Stage one

Here is a list of the basic tools and their approximate cost:

16-ounce claw hammer ($10)
Nail set ($4)
Ripsaw ($9)
Crosscut saw ($9)
Power saw ($20)
Brace/assorted wood bits ($15)
Screwdriver set ($7)
Block plane ($10)
Push drill and bits ($8)
Combination square ($6)
25-foot tape measure ($6)
¼-inch electric drill ($20)
Power sander ($10)
Pliers ($2)
Combination rasp ($3)
C-Clamps ($2 each)

Stage two

When your budget permits, add these tools to your basic set:

16-ounce ripping hammer ($10)
Tack hammer ($5)
Cabinet saw and miter box outfit ($15)
Hacksaw frame and blades ($6)
Keyhole saw ($6)
Razor knife ($1)
Bench vise (wood or metal working) ($20)
Level ($12)
Chisel set ($12)
Whetstone ($6)

Stage three

Fill out your collection with these tools:

Rubber or wooden mallet ($6)
Ball peen hammer ($7)
Coping saw ($4)
Folding rule ($6)
Marking gauge ($4)
Jack plane ($10)
Try square ($6)
Framing square ($10)
Dividers ($3)
Bradawl ($3)
Stapler ($15)
Flat bastard file ($5)
Mill bastard file ($5)
File cleaning card ($2)
Sliding T bevel ($5)

Not necessary, but nice to have

As projects increase, so will your tool inventory. And there is always Christmas, Mother's or Father's Day, and your birthday, when the family wants to know what tools to buy for you. Here are some suggestions to make your workshop complete:

Adjustable wrenches
Open-end and box-end wrenches
Socket set
2 (10-inch) pipe wrenches
Needle-nose pliers

Vise-grip pliers
Electrician's pliers
Wire strippers
Tinsnips
An extension cord
Expansion bit
Hole cutters for your power drill
Depth gauge for bits
Special screwdrivers (Allen wrenches, screw-holding drivers, offset drivers)
Masonry hammer
Star drills
Cold chisels
Propane torch outfit
Rabbet plane
Hand screw clamp
Metal files

Necessary accessories

Besides the basic tools, you will also need a selection of basic materials and accessories. These items include:

Sandpaper
Steel wool
All-purpose glue
An assortment of nails
An assortment of screws
Carriage bolts and fasteners
Household oil
Paste wax
Masking tape
Wiping cloths
All-purpose solvent

Hammer how-to

Of all the basic tools you need, hammers are number one. They are made in a variety of types and sizes, and have varying degrees of hardness and different configurations for special purposes. For example, a rubber hammer or wooden mallet is used to strike a chisel; a claw or rip hammer is used to hit a nail; and a masonry hammer is used to chip away at concrete and stone.

There are a couple of safety rules for the hammer that you should always follow:

1. Never use one hammer to hit another.
2. Always wear safety glasses when striking a hammer against a piece of metal or a tool, such as a star drill or cold chisel.
3. Use the correct hammer for the job. The face of the hammer should never be smaller than the tool that you will strike. And, of course, do not use a tack hammer to drive a spike, or a sledge hammer to drive a tack.
4. Never swing a hammer with a loose, broken, or damaged handle, because the handle may fly off.
5. If the face of the hammer is pitted or worn, try regrinding the face into its original shape. If you do not succeed, throw the hammer away and buy a new one.

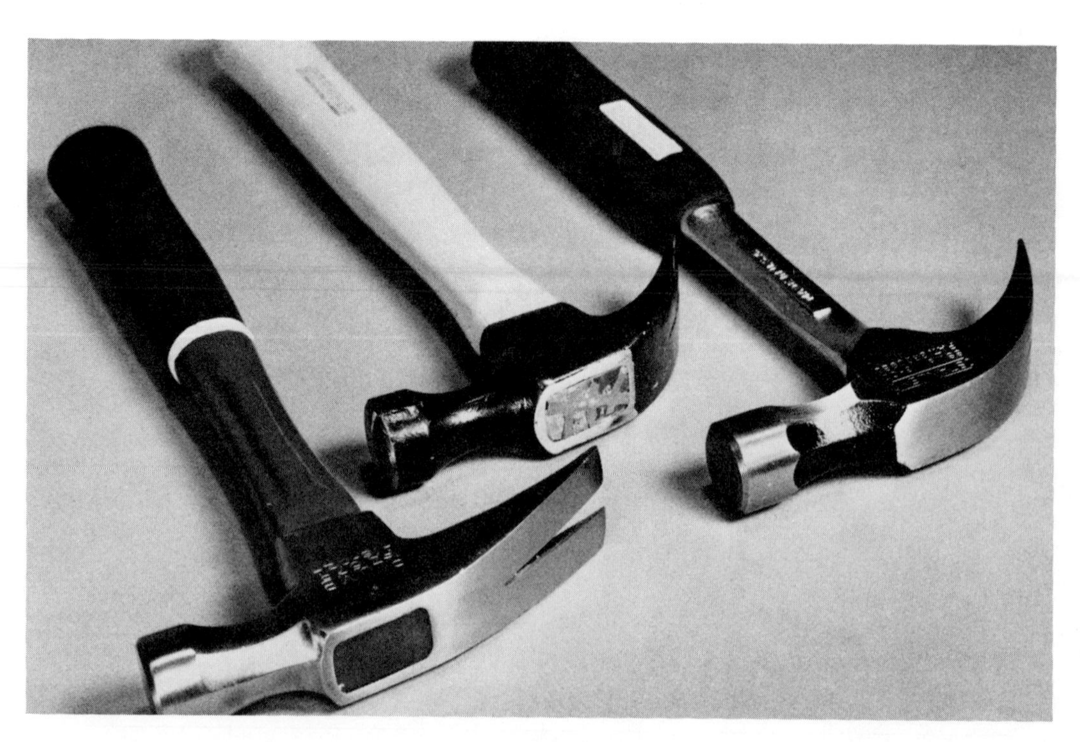

Quality hammers are easy to spot in a store; the faces of these hammers are well machined. Also look at the machining around the claws. When the machining is smooth, it is easy to slip the claws under a nail. Claws on a good hammer will pull a nail even if it doesn't have a head. Hammer handles may be made of hickory, solid steel, fiberglass, or tubular steel. The tubular, solid steel, and fiberglass models usually have a rubber grip.

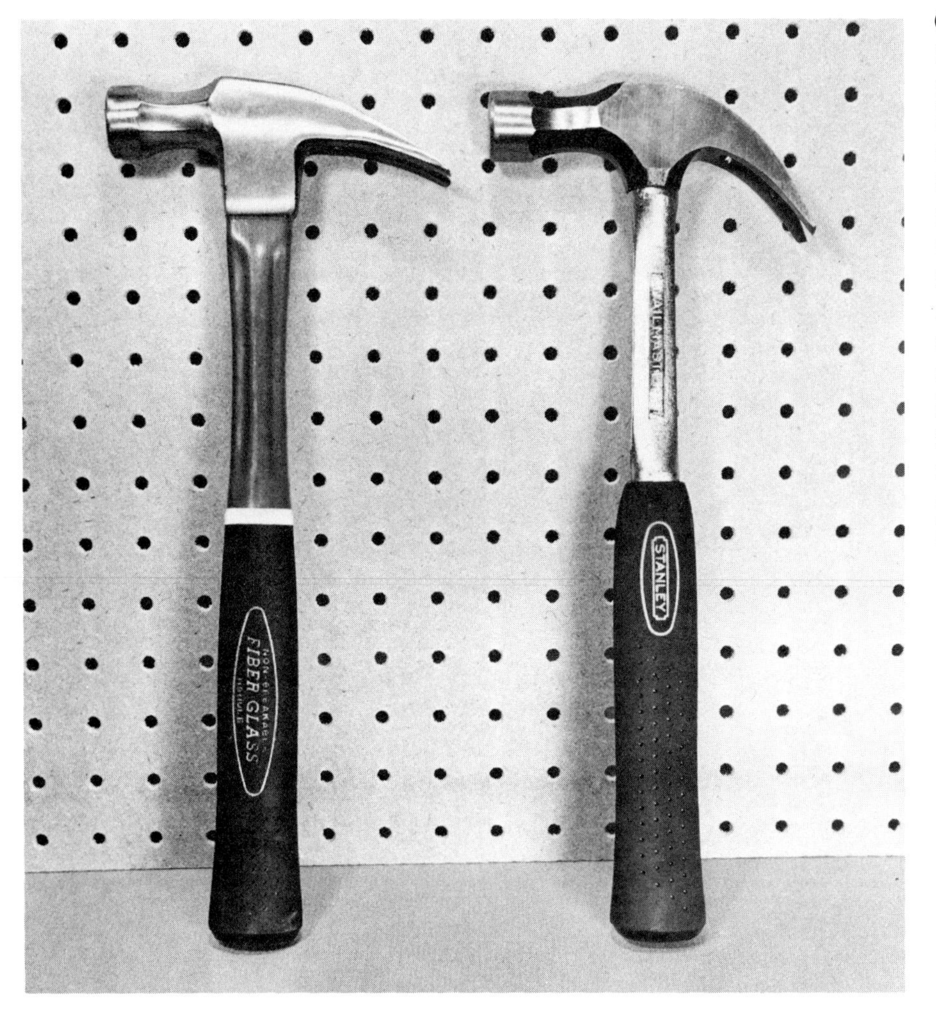

Common nail hammers are the ripping claw (left) and the curved claw. These hammers are designed for driving nails and nail sets; the claws are made for pulling nails and ripping woodwork. As a general rule, these hammers should not be used to drive hardened steel-cut and masonry nails; use a heavy hammer instead. Also, as a general rule, the hammers should not be used to strike cold chisels and other case-hardened steel tools.

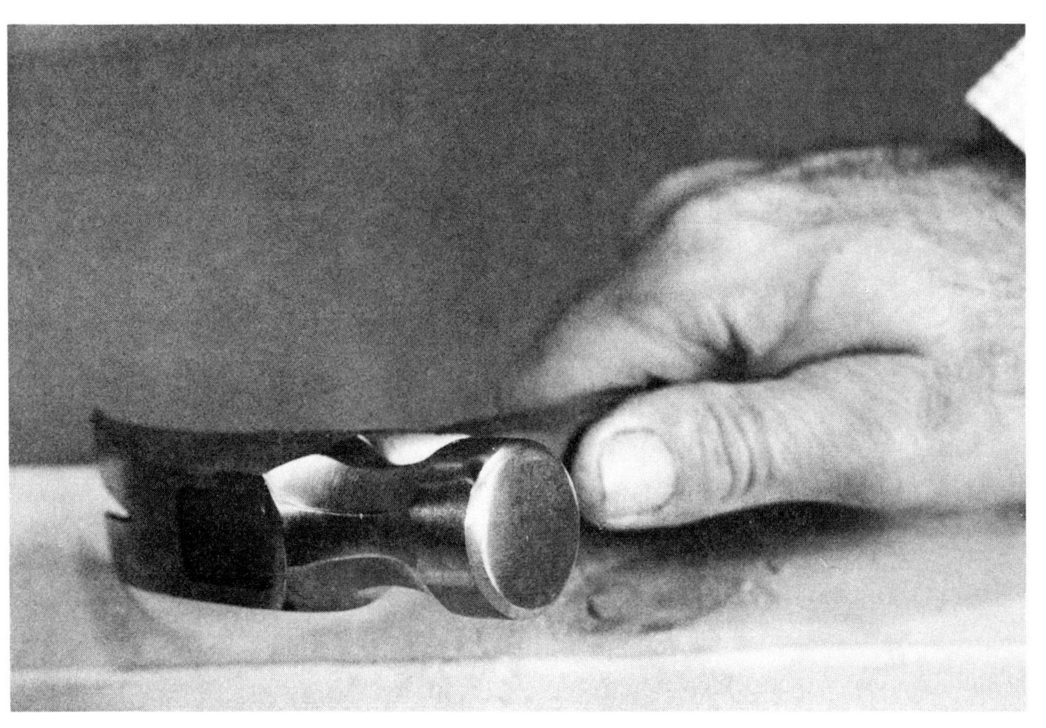

The face of a quality hammer is slightly curved at the poll (broad or flat end). With this design, a nail can be driven into wood without leaving hammer tracks on the wood. You may buy nail hammers in several weights. A 16-ounce hammer is best for the heavy work, when you'll be swinging the hammer quite a bit. A 13-ounce hammer is tops for light work.

Hold a hammer toward the end of the handle when you drive nails. This gives you more accuracy. Always strike the hammer with a square blow and have the face parallel to the surface you are striking.

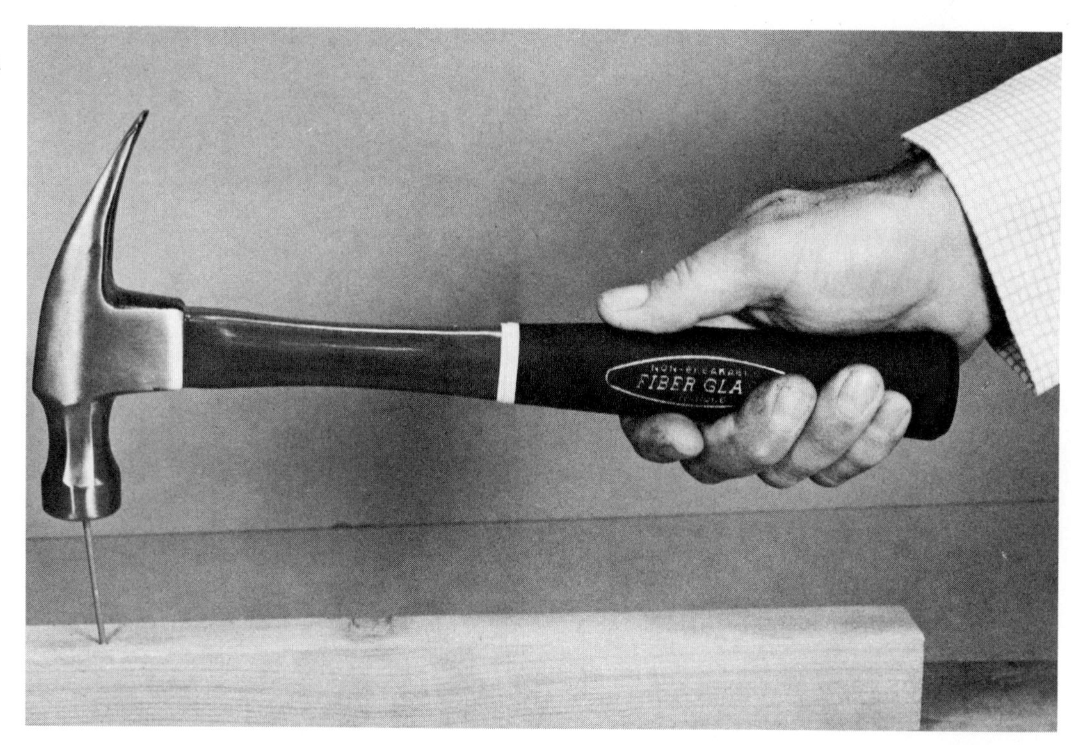

Choke up on the hammer handle (foreground) to start a nail; then, grip down on the handle to finish the job. To start a nail, tap it lightly into the stock (material). If splitting is a problem, hit the point of the nail with the hammer to blunt it, or drill a pilot hole for the nail. These tricks are especially useful when driving nails into wood such as maple or oak. Once the nail has started, hit it with even blows, and finish with light, even strokes.

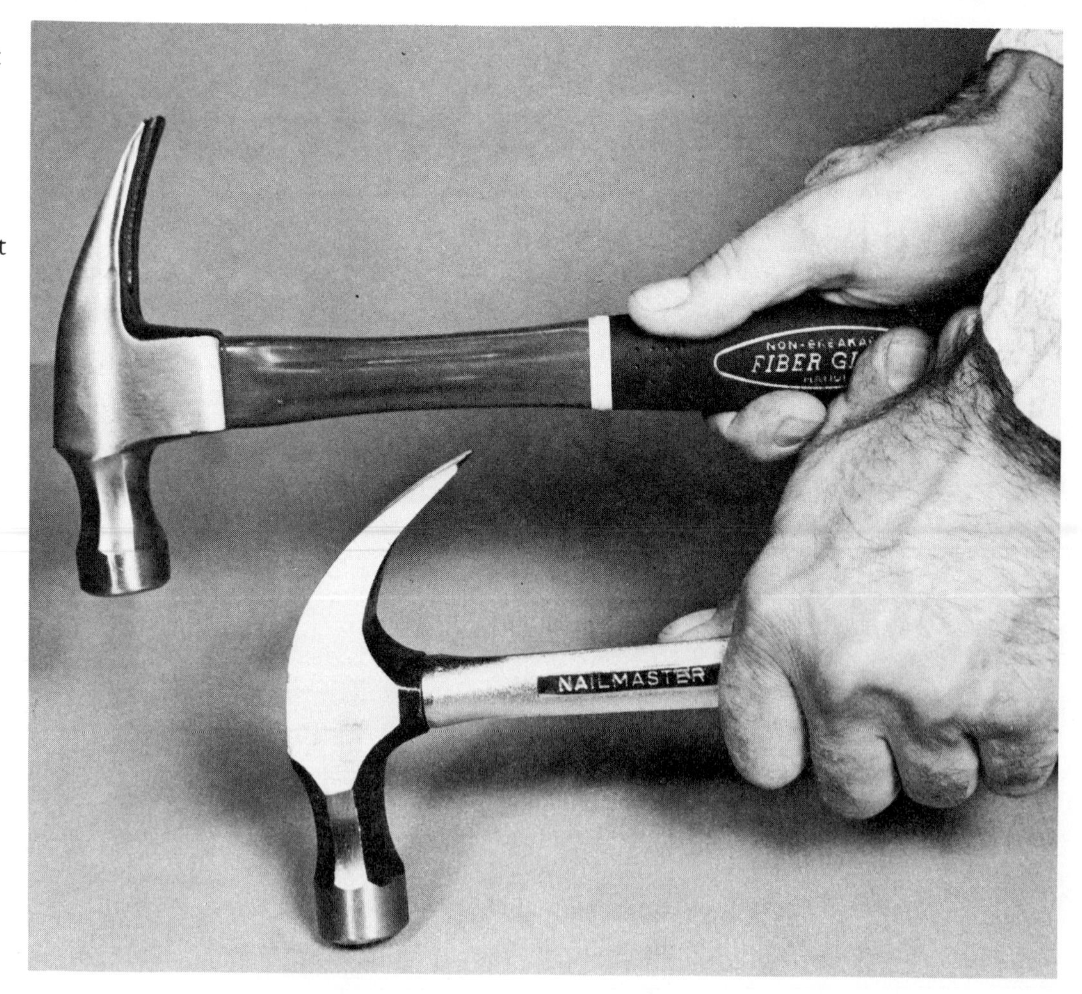

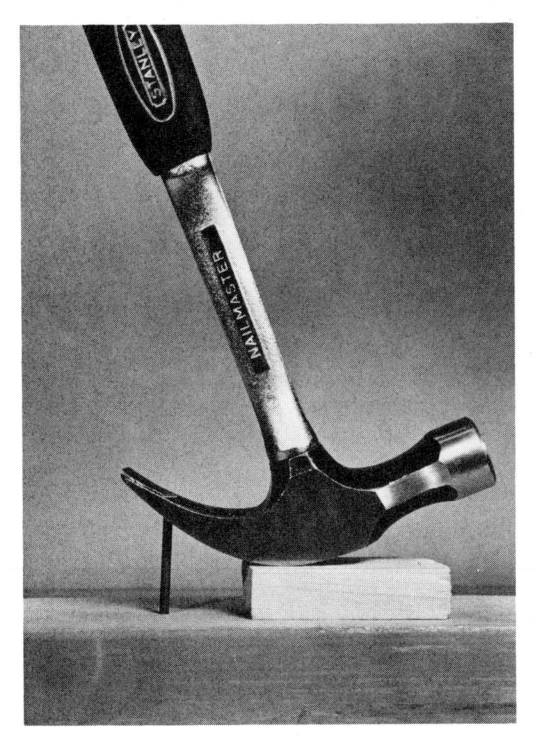

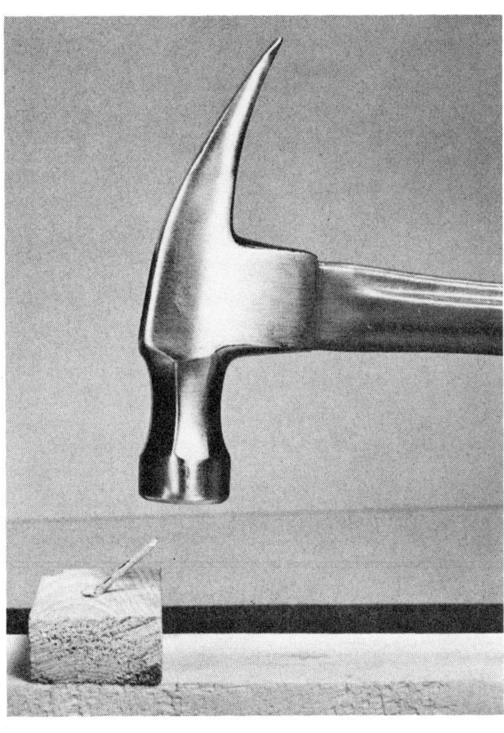

Far left:
Obtain more leverage when pulling a nail by inserting a small pry block of wood under the head of the hammer. Do this after the nail has been pulled out of the stock about an inch or so. For a really tough nail, use a pry bar with a nail puller. Otherwise, you may damage the hammer's claws or the wood. Or, you may break the handle of the hammer, if it is wooden.

Left:
Clinch a nail by hitting the tip of the nail over at a slight angle. Then lightly hit the nail to bend it downward into the stock, using light taps instead of heavy strokes. When the nail is on the surface of the stock, give the point a sharp tap, driving the point into the wood. A nail should be clinched with the grain of the wood, for holding power, whenever possible to do so.

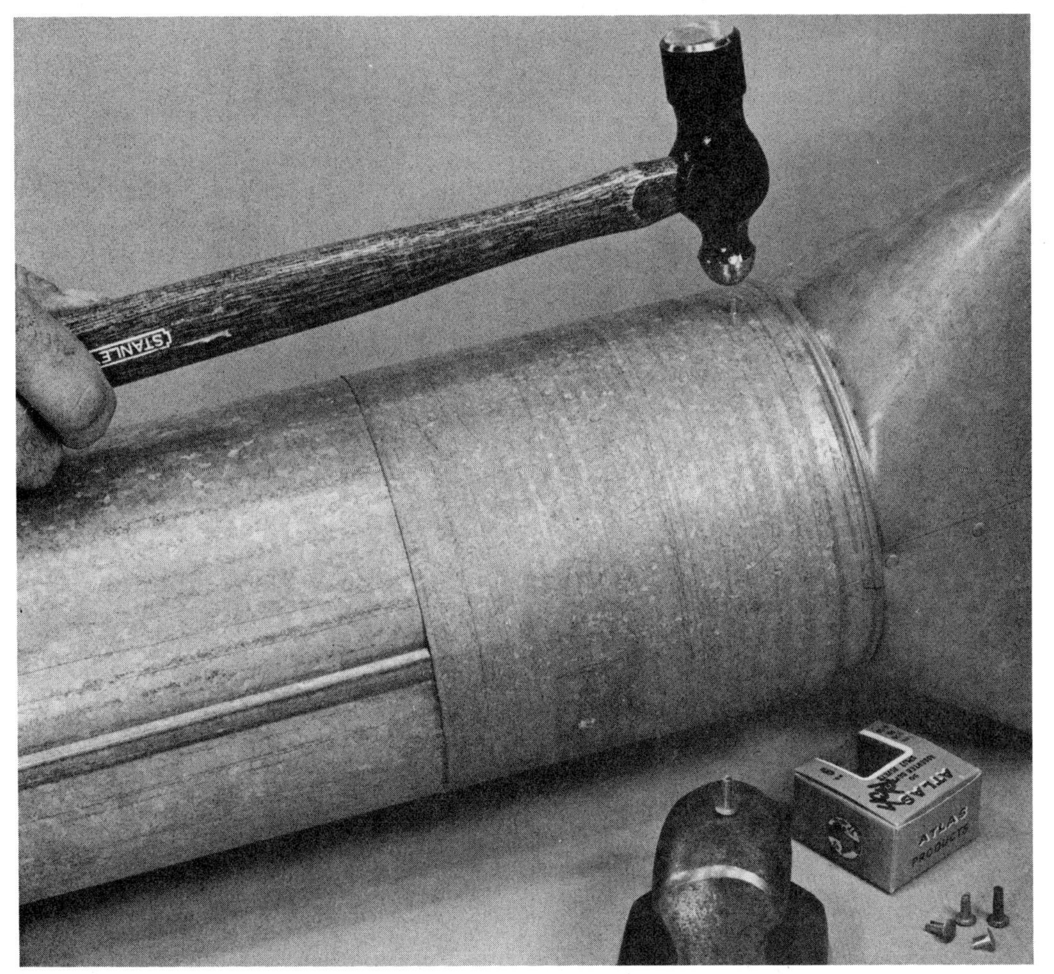

Ball peen hammers are designed with a slightly curved striking face. On the other end is a round, ball-shaped peen. These hammers are commonly used for riveting. They may also be used for striking cold chisels, and are handy for shaping and straightening metal material. Ball peen hammers are available in several weights—up to 3 pounds.

Right:
Quality tack hammers are magnetized on one end of the head. This holds the tack in place so that you may start it into the material. Some tack hammers have a long thin claw for pulling tacks out of corners and tight quarters. Forged from high-grade steel, tack hammers weigh from 5 to 8 ounces. The 8-ounce model is recommended for most ordinary tacking jobs.

Far right:
Starting a nail in tight quarters can be a problem. Try sticking the nail through a piece of light cardboard or newspaper, then hitting the nail with a hammer. After the nail has started into the stock, you can pull the paper holder loose. This trick is especially useful for starting small brads and tacks if you don't have a tack hammer with a magnetized head.

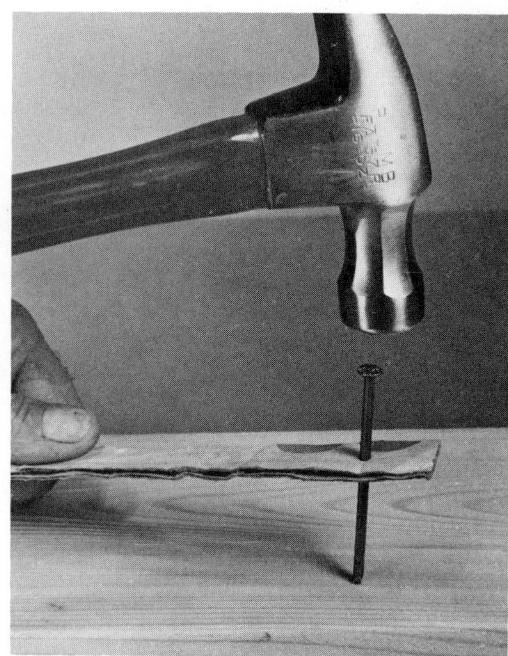

Nail sets countersink nail heads below the surface of the stock. There are several size nail sets available to fit different size finishing nails. Always fit the points of the nail sets to the nails. If the nail sets are too large, they will tend to damage the stock; if the nail sets are too small, the nails cannot be countersunk properly. Always countersink the nails (when the job calls for it) about ⅛ inch below the top surface of the stock. The holes may be filled with wood putty to cover them.

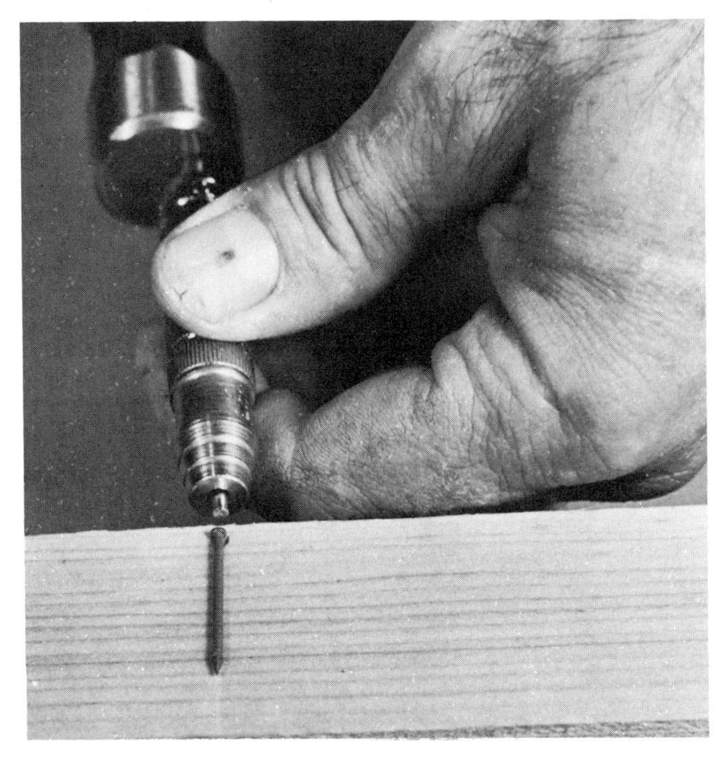

A new type of nail set has a driving pin inside a metal housing. Square the nail set over the nail before driving it flush into the wood. Then tap the top of the metal housing, which, in turn, strikes the pin and the nail, driving it below the surface. The advantage of this tool is its ability to countersink the nail without bending it, or damaging the face of the stock.

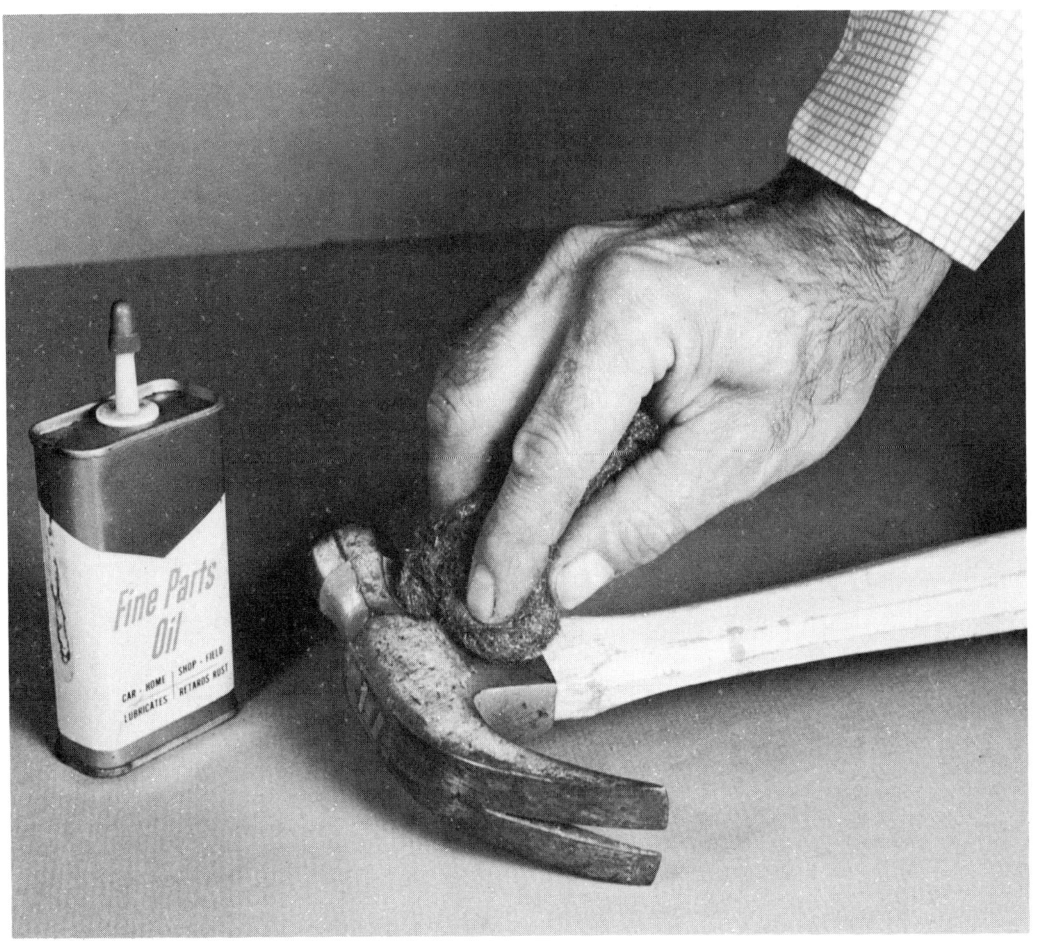

Keep a hammer rust-free by giving it an occasional buffing with fine steel wool. Use light household oil on the steel wool. When the face and poll of your hammer becomes pitted and worn, try reshaping the face with a grinder or metal file. If the face is really damaged on any hammer, throw the hammer away and buy a new one. A damaged hammer can be dangerous.

Saw how-to

The basic hand saws used in carpentry are the crosscut saw and the ripsaw. If your budget permits, buy both at the outset; if not, buy the crosscut saw first, since it may be used for most sawing jobs. Then add the ripsaw.

A crosscut saw and a ripsaw are named because of the way the teeth are set and the number of teeth per inch along the blade. A crosscut saw has from 6 to 12 teeth per inch. The teeth are beveled (angled) on the front and rear edges and are "set" alternately, one tooth "bent" out on one side of the blade, the next "bent" out on the other side of the blade. A ripsaw has from 4 to 5½ teeth per inch. Each tooth is sharpened across the face and rear edges, and they are also set alternately and outwardly.

A *ripsaw* is used to saw lumber with the grain of the wood. Hold the saw at about a 60 degree angle without pushing the saw heavily into the wood; the weight of the saw is enough to make the proper cut. If you have to bear down, the saw probably needs sharpening—a job for a professional. Always work about 1/16 inch from the scrap side of the marked cutting line.

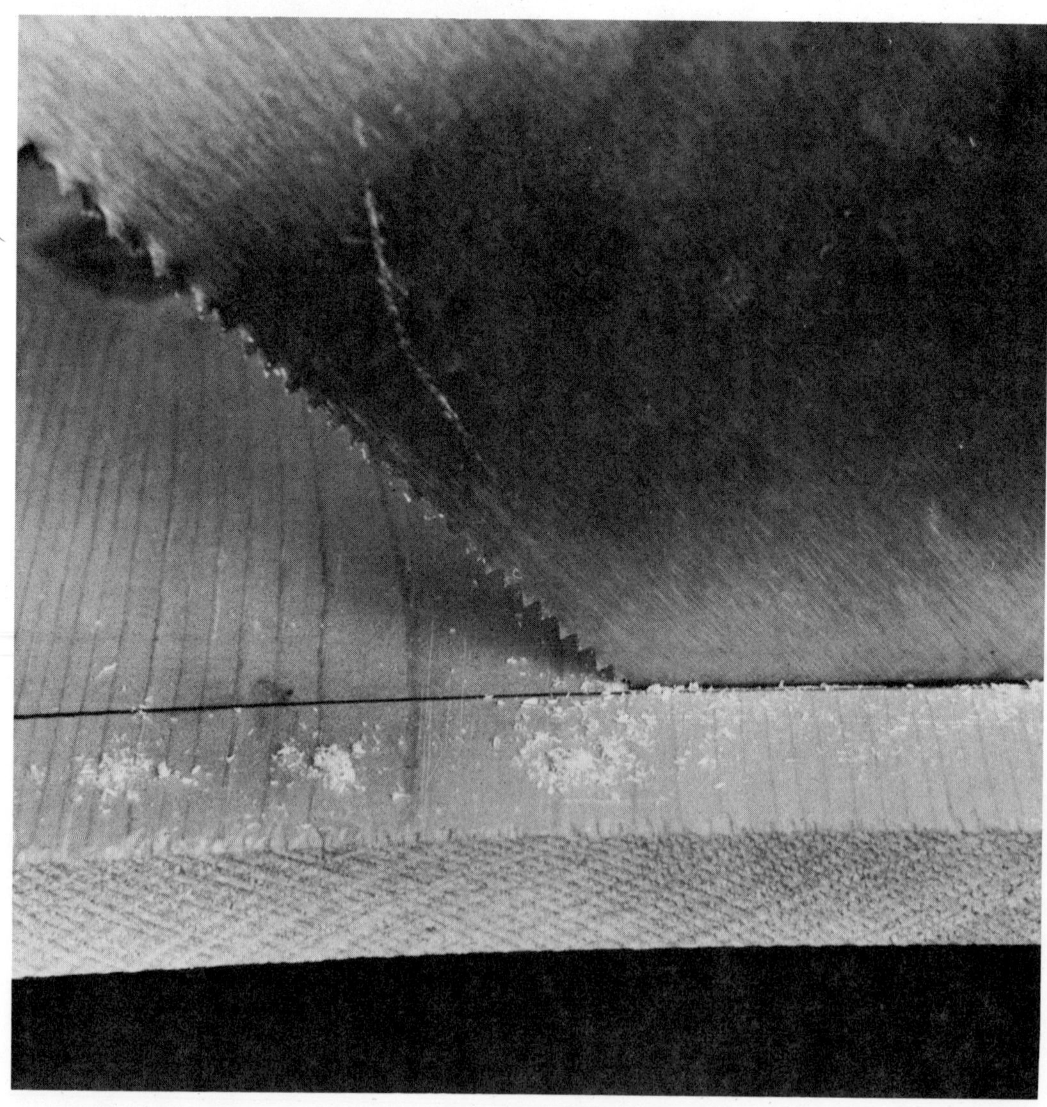

A *crosscut saw* is used to saw lumber across the grain, although it can also be used as a ripsaw. Hold the saw at about a 45 degree angle to the wood, and don't force the cut, or the blade will twist and leave the cutting mark.

A portable electric saw can save time when you need to cut plywood, hardwood, and lumber. Buy a saw with a 7¼-inch blade; this size will cut 2 inch thick lumber easily. Also, the saw may be set to make angle cuts. It is necessary to own 2 blade types: a rip blade (has larger teeth than a combination blade) and a combination blade (teeth are not set so far apart as the rip blade teeth). If you want to cut masonry, buy a carbide-tip blade.

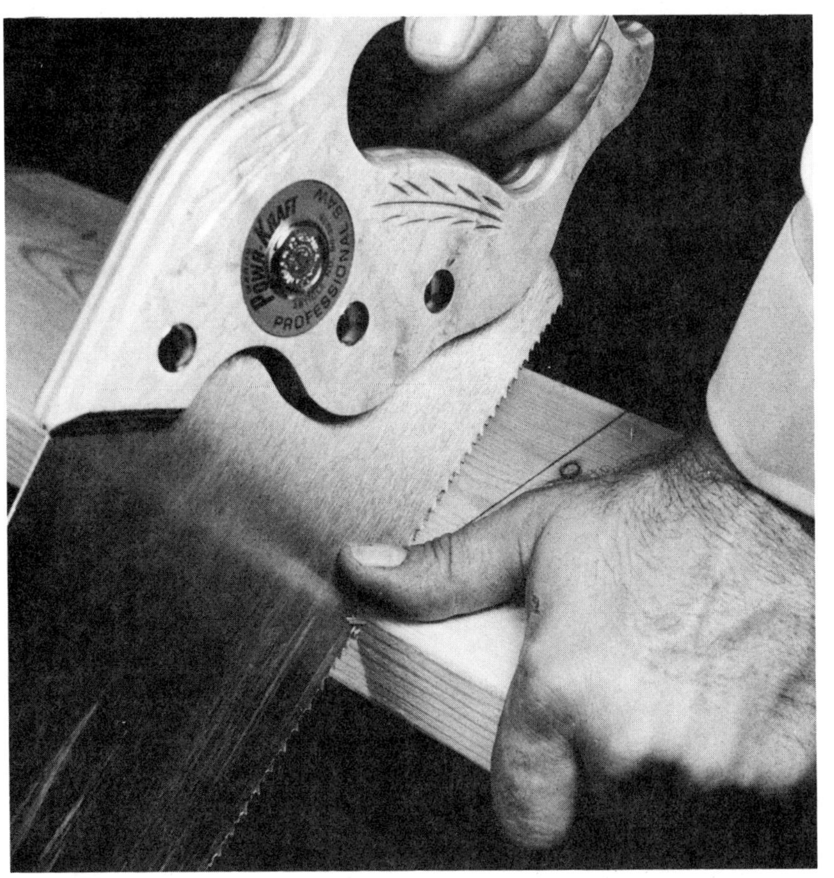

Start the saw into the wood with a backward pull. Go easy; you just want to notch the wood to guide the saw. Use a knuckle or finger of your free hand to start the saw at the mark. Keep the guide finger in place for a couple of strokes, or until the saw is properly started into the wood. The sawing action is from the shoulder rather than the wrist or hand; let your shoulder swing in a normal arc. This will produce a rocking motion that will guide the saw along the cutting mark.

The saw won't jam in the kerf (the slot or cut that the saw makes) if it is kept open with a block or wedge of scrap wood, or a spike. When sawing in green wood or wood that is sappy, coat the saw blade with a light household oil or paraffin. The paraffin trick should be used if the wood is going to be finished with a stain.

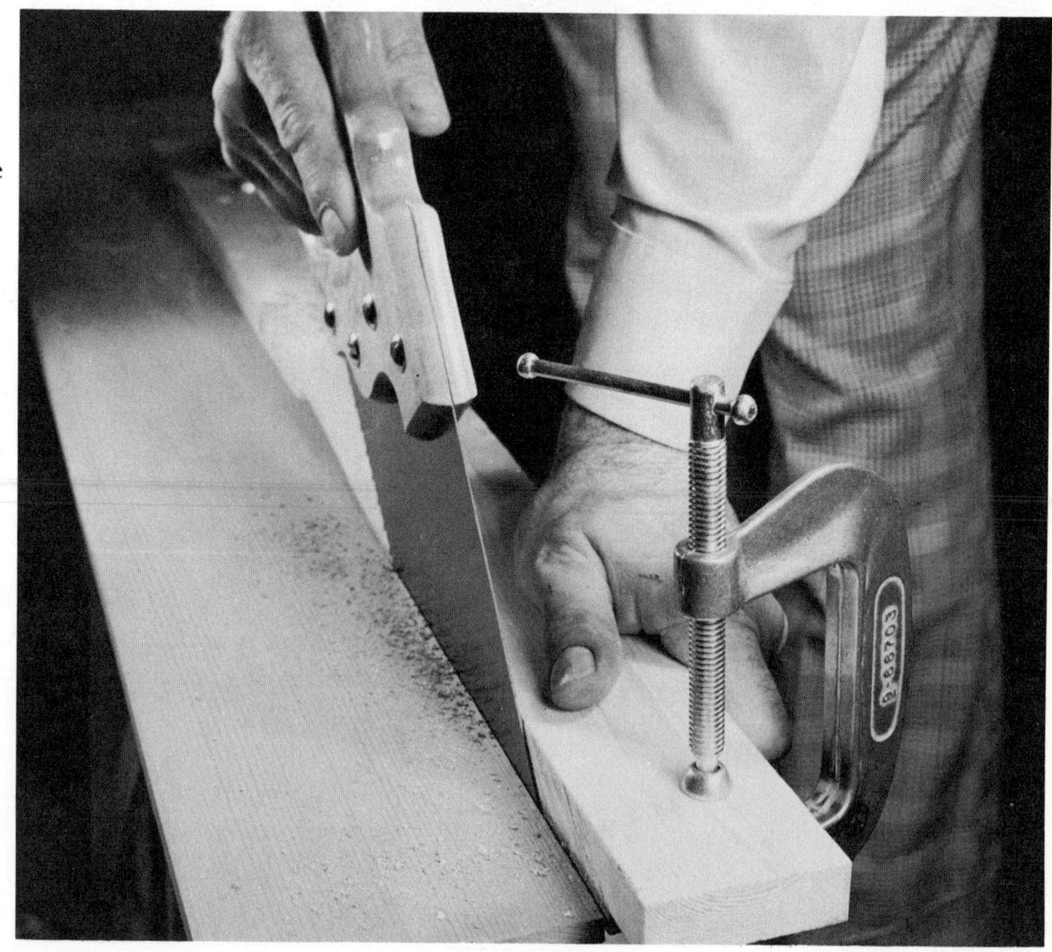

A straight ripping guide can be fabricated by clamping a straight, square piece of wood to the stock. However, the stock should be marked for the cut. If the saw wanders from the line when you're using it, go back to the spot where it wandered and start again. Don't try to twist the saw back onto the line, or the cut will be out of square. When you're ripping long pieces of lumber, check the cut frequently to make sure the saw is square in the kerf. Keep your crosscut and ripsaws clean and polished by buffing them with fine steel wool and a coating of household oil. If the handle is loose, try to tighten it with a screwdriver; if the handle is broken, buy and install a new one.

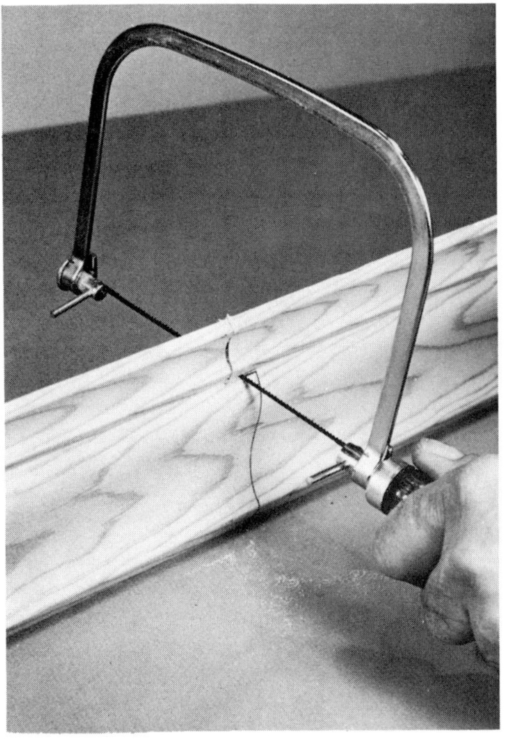

Far left:
A *coping saw* has a rectangular frame with a tension adjustment to hold the blade taut. Turn the handle of the saw to apply tension to the blade. Blades have 17 points or teeth to the inch which produce fairly fine cuts. The big advantage of the coping saw is its ability to turn at almost any angle during the sawing process. To set the blade at the desired angle, twist the 2 metal pins where the blade connects to the saw frame.

Left:
A *hacksaw* is used for cutting metal. The blades cut on the forward stroke and must be mounted in the frame so that the teeth slant forward. Like the blades on a coping saw, hacksaw blades are under tension, which is created by twisting the handle.

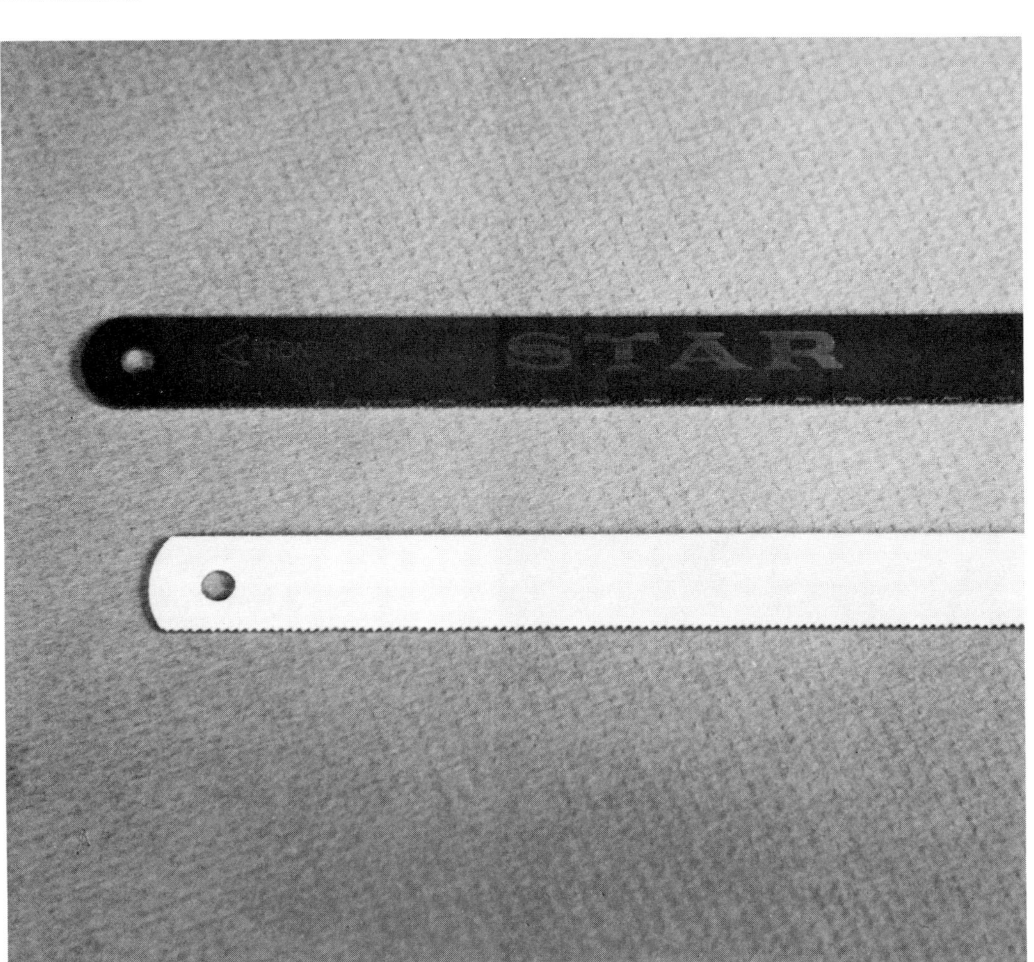

Hacksaw blades have from 18 to 32 points or teeth per inch, and you can buy special blades for cutting such materials as plastic and glass. The fine-toothed blades produce the smoothest cuts. Hacksaws may be used to cut copper tubing, steel, aluminum, and armored electrical cable. If the blade tends to skip on the metal, use 2 blades in the frame—one with the teeth pointed forward, the other with the teeth pointed backward. Of course, the saw kerf will be wider.

Sawing in cramped quarters with a hacksaw is easy; just turn the blade over on the frame after you assemble the saw "around" the work. You can buy a hacksaw with a keyhole saw configuration which is ideal for cutting a hole in metal or for using in tight quarters.

If you have a damaged screw slot, back the screw out of the stock as far as possible and cut a new slot with a hacksaw blade. Other hacksaw tips: On tempered steel and pipes, notch the cutting mark with a file. This anchors the blade so that it doesn't skip or wander. When cutting thin metals, such as aluminum or copper sheet, clamp the metal between 2 pieces of scrap lumber. Saw through the lumber and metal at the same time to prevent the metal from bending and twisting. If the hacksaw frame won't slip into an opening, remove the blade from the frame and wrap one end of it with masking tape for a handle.

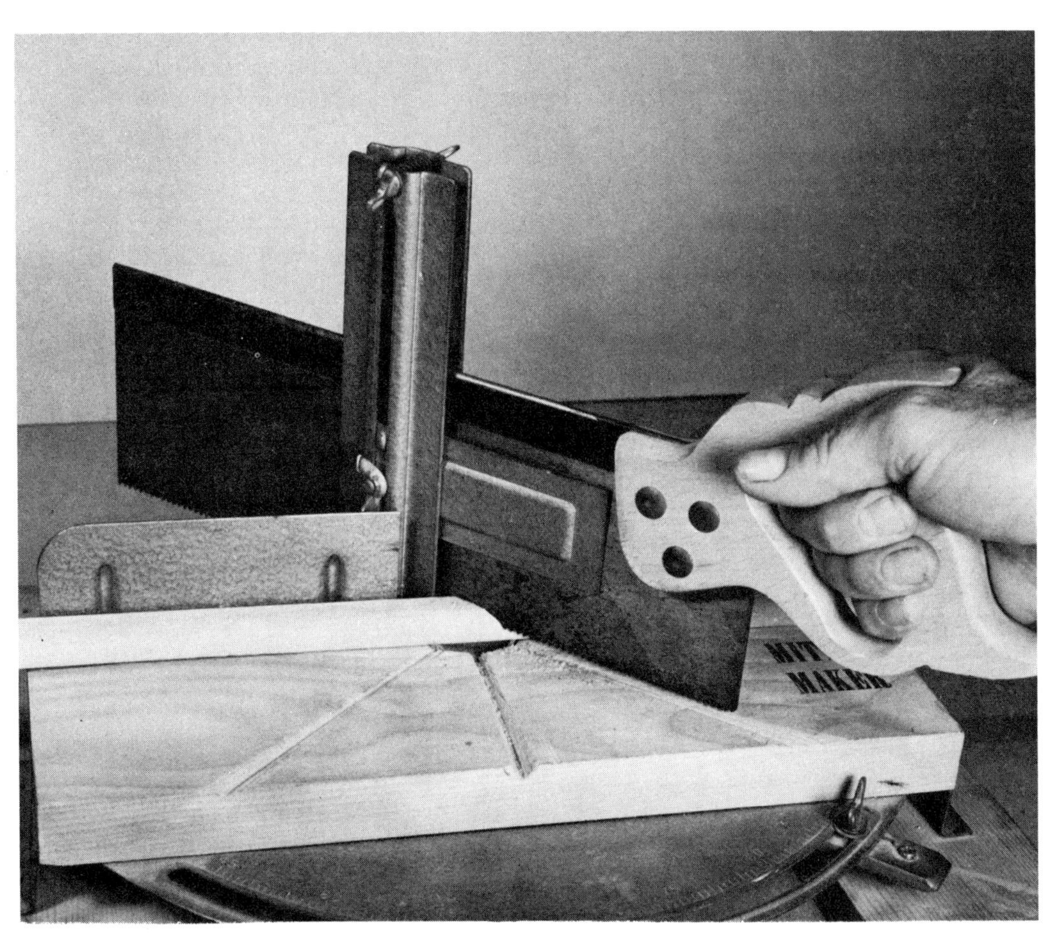

A *backsaw* is used for cabinet work and in miter boxes (a device for guiding a handsaw at the proper angle) It has from 18 to 32 points per inch, and the teeth produce a very fine cut. Backsaws are manufactured in several lengths, from 12 to about 30 inches. When you use a miter box to cut an angle or to make a square cut, lock the saw holder so that it can't slip and throw the saw off the line. Also, when you cut miters, cut about 1 degree less than intended, 44 degrees instead of 45 degrees, for example; this will make the joint tighter when fastened.

Far left:
If you do not have a miter box, cut fairly accurate angles by overlapping both pieces of stock and sawing through them at the same time. The cuts will match. Use a fine-toothed backsaw or coping saw for this.

Left:
A *keyhole saw* has a pistol-like grip, and is used to cut circles and angles in wood or metal. The blades have from 8 to 24 points per inch and are tapered to slip easily into a starting hole.

Chisel how-to

A chisel cut is made with the grain of the wood. If you push the chisel in the opposite direction, away from the grain, you may split the wood. When making chisel cuts, don't bite off more than the chisel can chew. Keep the cuts shallow; go easy.

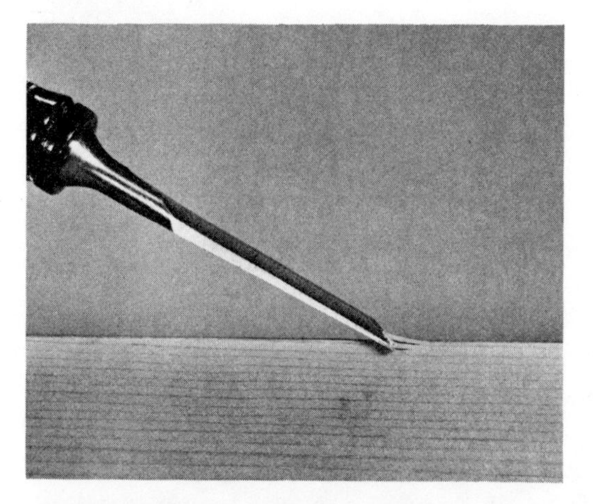

Chisels are in the same family as knives; they whittle, smooth and make shallow cuts, and trim off little bits of wood that are in the way.

A chisel set is a good basic tool to own. The set includes four sizes: ¼-, ½-, ¾-, and 1-inch widths, and this selection will handle almost any job. Later, you may want to add other sizes.

Chisels must be kept razor-blade sharp. A dull chisel is dangerous; it can cut too hard into a delicate area, causing a riblike marking along the wood. A bad cut will cost you the price of the wood, and injure fingers.

Push the chisel sideways along the cut. The blade has a shearing action. The pressure is in the palm of your hand, with a finger guiding the cut and applying a slight downward pressure. For very hard woods, use both hands for shaving: One hand supplies the forward thrust, the other a downward pressure on the blade at about the halfway point.

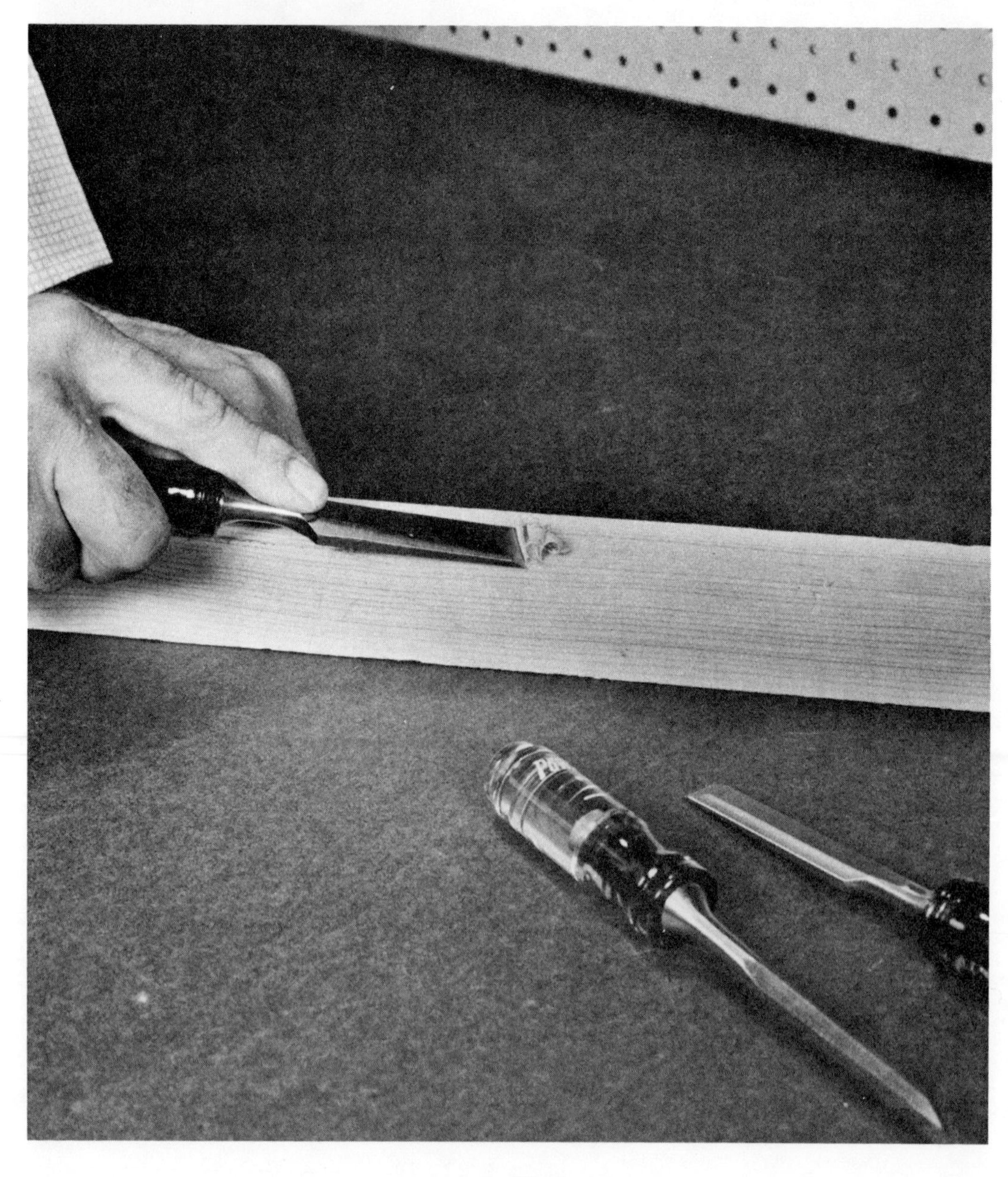

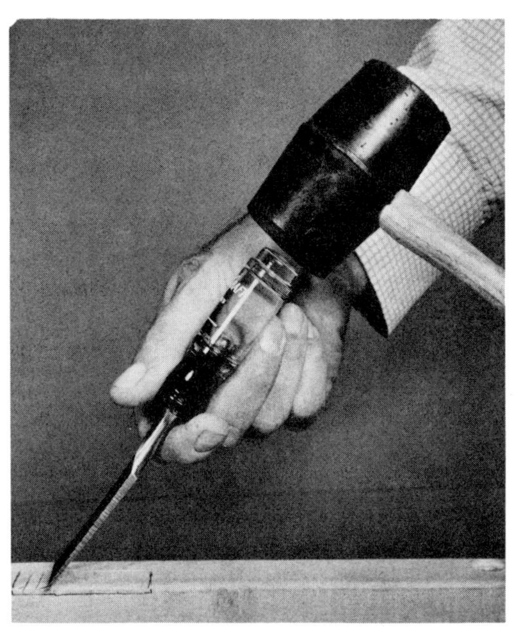

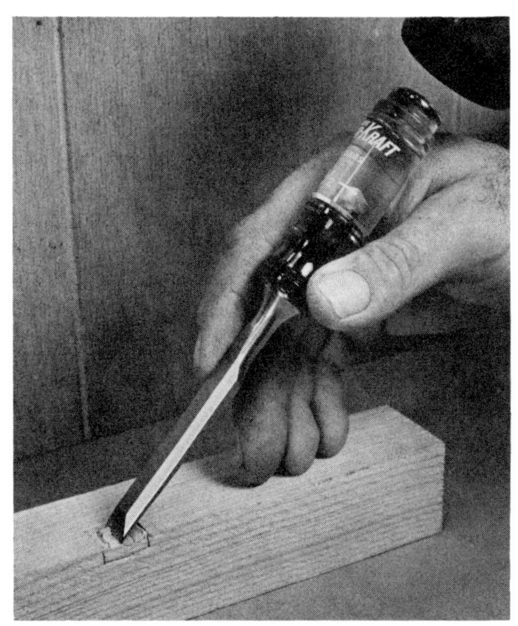

Far left:
Mortise (recess) cuts with a chisel in which the chisel will penetrate the wood to ⅛-inch depth or so, require a rubber hammer or wooden mallet to strike the handle of the chisel. Using a metal hammer will damage the handle of the chisel. Hit the handle of the chisel squarely with the rubber hammer.

Left:
Keep the bevel of the chisel blade up for most cuts. If the chisel cannot lie flat on the stock with the bevel up for a smoothing cut, hold the chisel flat-side up at a 20 degree angle or so. The taper on the blade will help maintain this angle throughout the cut.

The primary purpose of a chisel is to cut mortises in edges of doors for hinges. You may also (at some time) have to recut an old mortise to unstick a door. To cut a new mortise, first outline the shape of the hinge on the edge of the door with a sharp pencil, as shown.

Right:
Outline the cutting marks with the chisel, tapping the flat side of the blade into the wood. Hammer the chisel to the depth you want the cut to be, usually about ⅛ inch, or the thickness of the hinge. Mark this depth on the side of the stock as a guideline.

Far right:
Make a series of chisel cuts within the boundaries of the cutting lines, as shown. These should be spaced about every ⅛ inch. Keep the tapered edge of the chisel down, and tap the chisel into the wood at about a 60 degree angle. Tap lightly: chisels cut deep. For the top veneer cuts in plywood, cut out the solid wood first, as in the frame of a door. Then trim away the veneer.

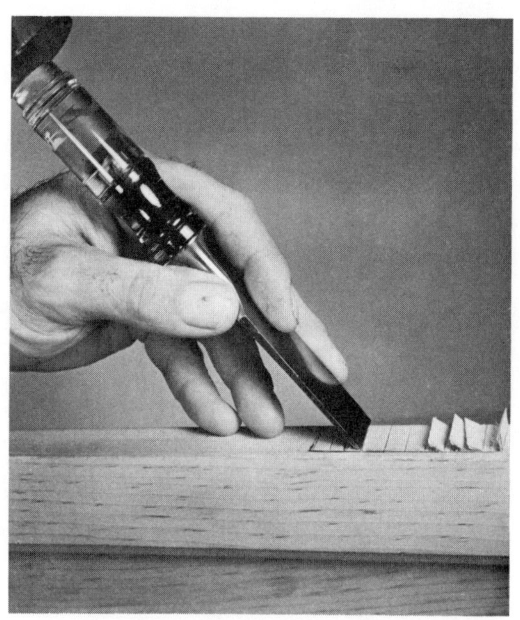

Smooth the mortise cut, breaking out the precut chips, with the taper of the chisel in a downward position. Shave the stock at a slight angle to get a smooth base cut; do not try to take too much wood off at a time. Make 2 shallow cuts, then fit the hinge into the mortise to check the depth, width, and length. Keep cutting and checking until the hinge fits properly into the cut. If you cut the mortise too deep, the hinge can be shimmed (filled in with a small piece of metal, wood, or other material) with cardboard.

Keeping your chisels sharp and in shape

Far left:
Keep your chisels sharp; this is a "must." Touch up the cutting edges before each project by running the edge (taper down) in a figure eight pattern across a whetstone or oilstone. Three or four turns around the figure eight are enough to whet the edge.

Left:
Nicks and mars on a chisel's cutting edge are sometimes difficult to remove by running the taper over an oilstone. Before you throw the chisel away, try reshaping the blade with a metal file or on a grindstone. File or grind the flat part perfectly smooth, after you have squared the cutting edge, removing the nicks. Then retaper the leading edge of the chisel with a file (or grinder), and whet the edge on a whetstone.

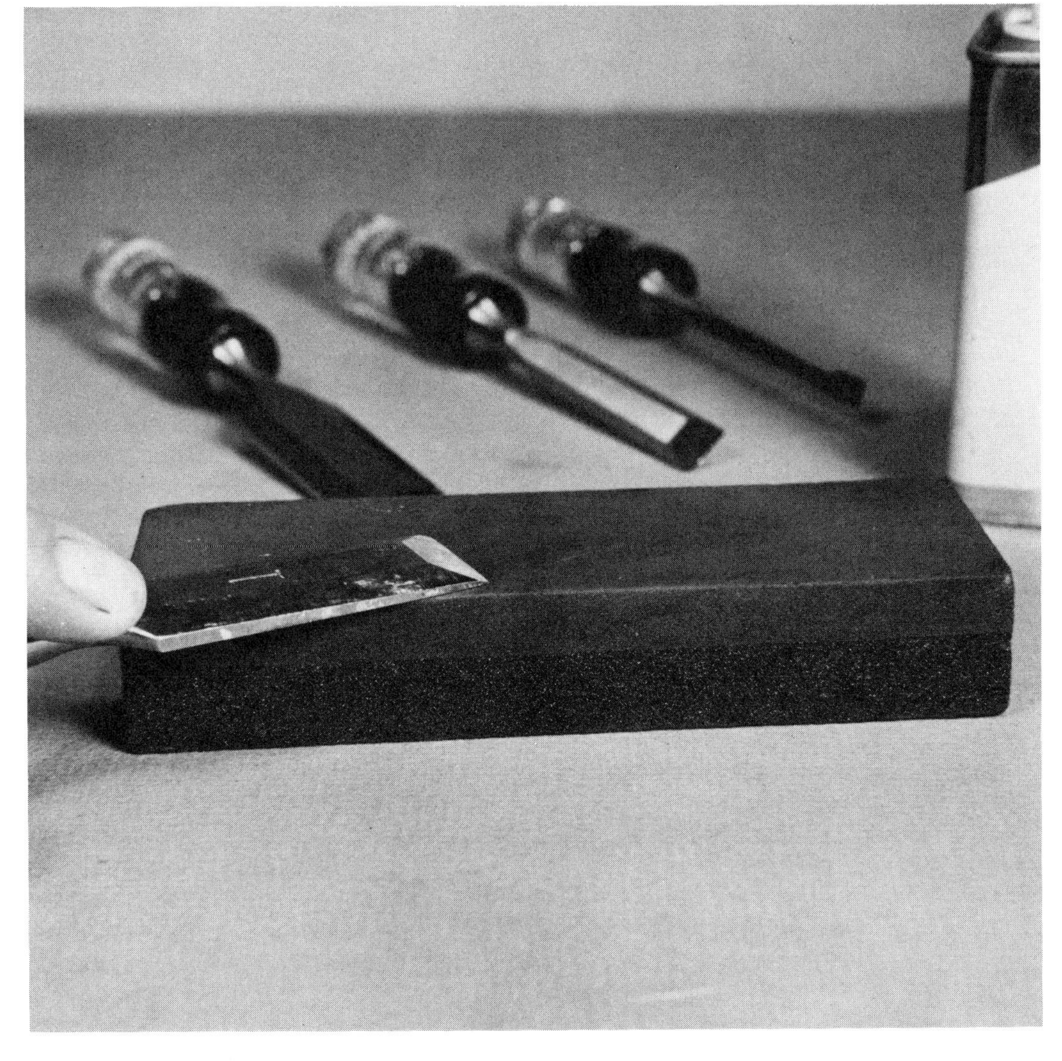

A *slight burr or ridge* may appear on the flat side of the chisel after you sharpen the tapered edge. If so, run the flat edge of the chisel, in a flat position, across the whetstone. This will remove the burr. You can test the sharpness of the chisel by running the blade slantwise through a piece of newspaper. The blade should cut the paper like a sharp razor blade.

Plane how-to

A plane forgives your cutting mistakes; it is like an eraser on a pencil, the reverse gear in a car, or a pal in court. If you cut a board and it isn't perfectly square, a plane can make it square. A plane can also smooth rough, cupped, and pitted lumber, which saves buying replacement boards.

The plane family includes block planes, jointer planes, smoothing planes, rabbet planes, cabinetmaker's planes, spokeshaves, and low-angle block planes.

The first plane you should buy is a jack plane; it is heavy, but short enough to handle most planing jobs. When your budget permits, buy a block plane, then a smoothing plane. Let the rest be given to you; they are not necessary, but nice to have.

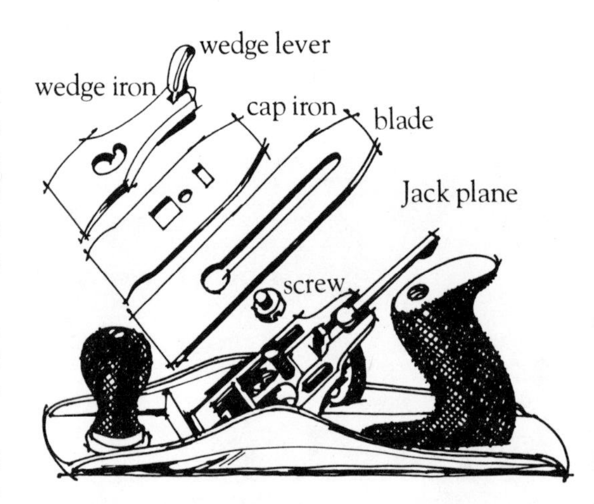

Like chisels, planes must be kept as sharp as a razor; a dull plane can ruin good lumber.

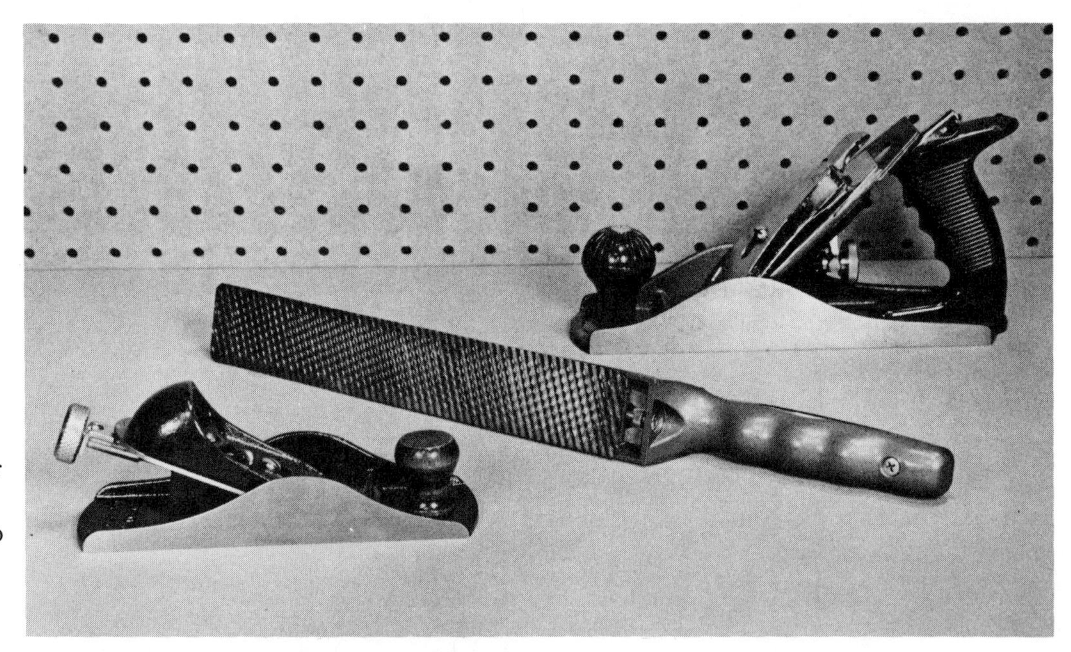

A plane selection for a basic workshop should include a block plane (left), a multibladed forming tool (center), and a jack plane. The multibladed forming tool is fairly inexpensive, and you can buy a potful of different blades for it since the blades are interchangeable. The blades include a file, regular plane, and convex plane for curved shapes. Also, a special blade is available for smoothing metal; a half-round blade may be used to smooth concave and compound curved material.

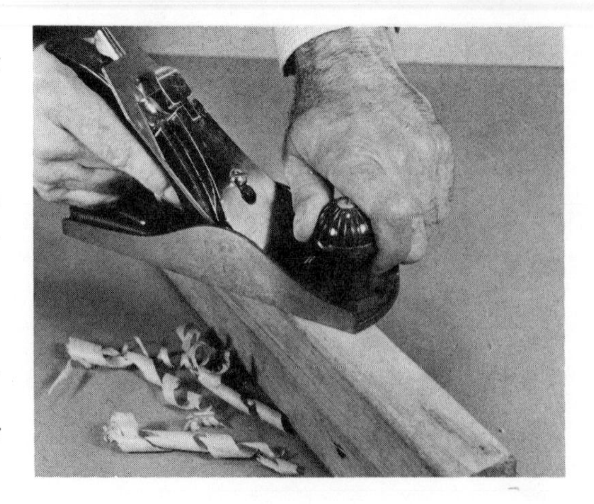

A plane has a shaving action similar to a chisel, razor, or pocketknife. Run it across the stock at a slight angle, as shown, for best results. Also, here's a trick you can use: Run a couple of pencil lines along the surface to be planed, then run the plane over the stock. The remaining pencil lines will show you whether the plane blade is high, low, or square to the surface.

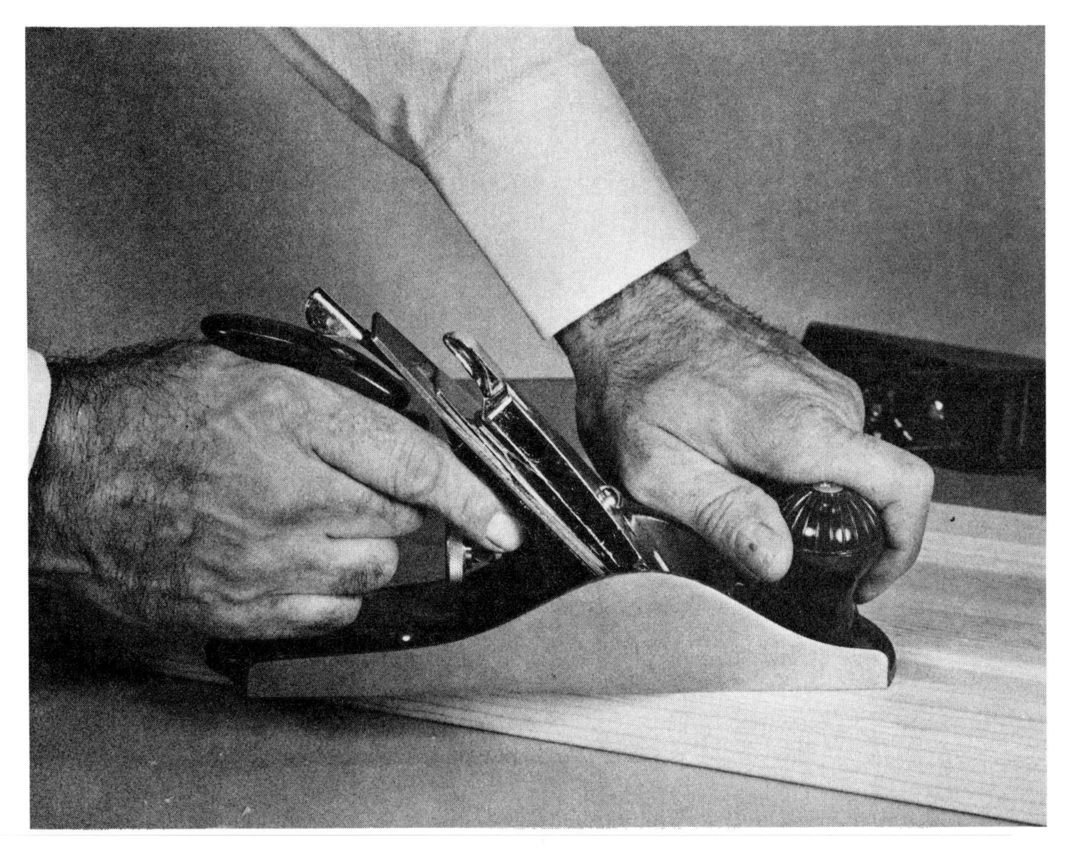

Apply the right pressure to the plane as you push it across the stock. At first, the pressure should be forward toward the front of the plane. During the middle of the cut, pressure should be applied to both the front and back handles. At the end of the stroke, pressure is applied to the back handle. You don't need much muscle. Go easy. If the shavings are not the same thickness or do not fall continuously, the blade may be dull.

A *multibladed plane* goes with the grain and across the grain, and can do the work of four tools—a file, rasp, plane, and sandpaper. It is difficult to make a mistake with this tool, and better yet, new blades are inexpensive and easy to replace by turning a couple of screws. This tool will handle wood or metal in almost any shape, as well as plastic material.

Right:
A *block plane* is designed for small, delicate planing jobs —mostly cabinet-type work. Use it for short, accurate cuts—here a chamfer (slight angle) on a piece of trim. For the best cut, apply pressure on the front of the plane with a finger. Apply the rest of the pressure to the top blade assembly with the palm of your hand. A block plane is especially useful when planing and smoothing the end grain of lumber.

Far right:
The blade of the plane should be parallel to the bed (bottom) of the plane. Do this by moving the lateral adjustment lever, which is located in back of the plane iron and the plane iron cap. You can tell if the blade is parallel by sighting the bed of the plane.

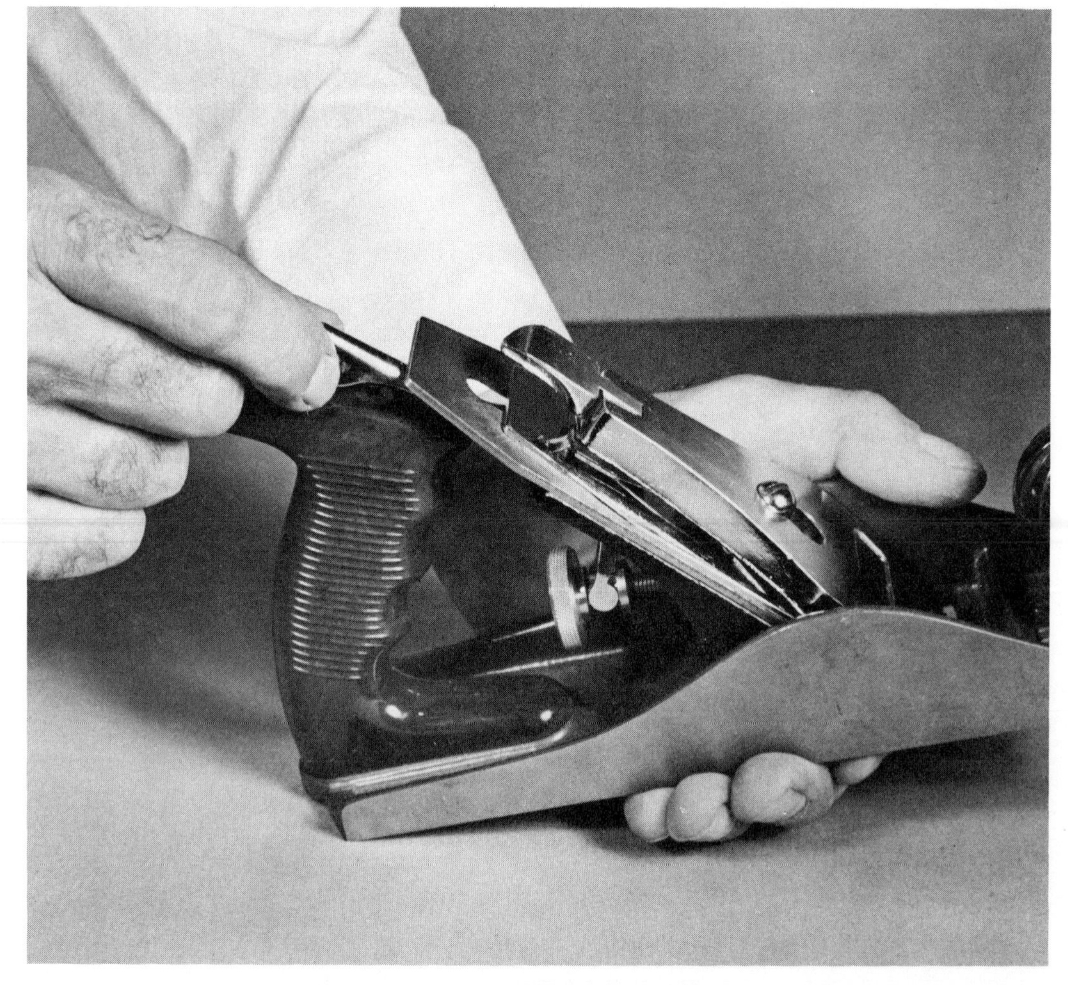

An *adjustment lever* controls the angle of the blade. In a normal, halfway position, the blade of the plane should be square with the bed. Flip the lever to one side or the other, and the angle of the blade will move to the bed (or bottom) of the plane.

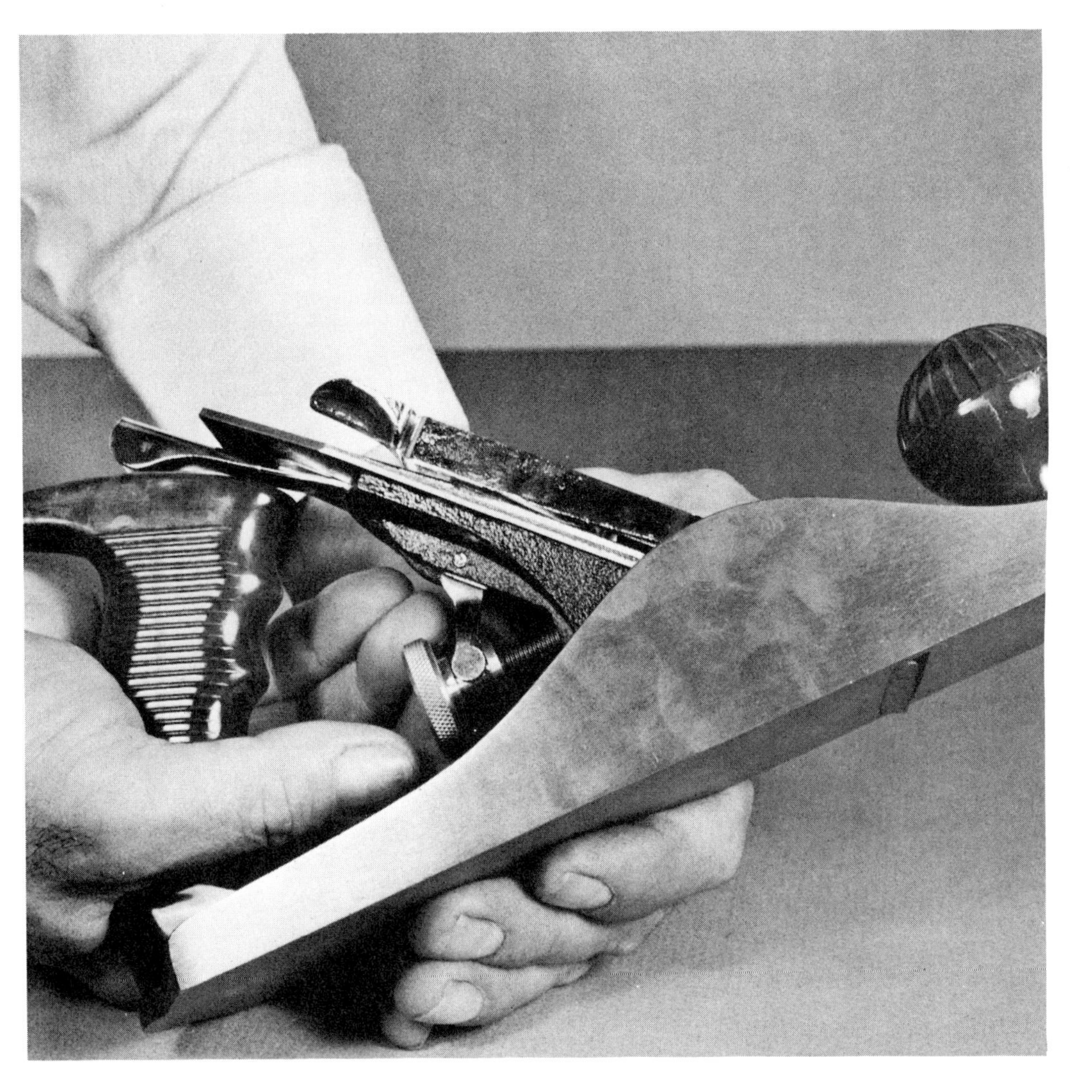

The screw nut in back of the plane blade assembly adjusts the depth of the cut of the plane. As you turn this screw, run your fingers lightly over the blade on the bed of the plane. This will help you determine how deep the blade adjustment should be. The blade should protrude just a tiny bit for a smooth cut; if you want a deeper cut, turn it down farther. Do not turn the blade down too far or the plane may dig into the stock. It is always best to test the depth of the cut on a piece of scrap before you run the plane over the good stock.

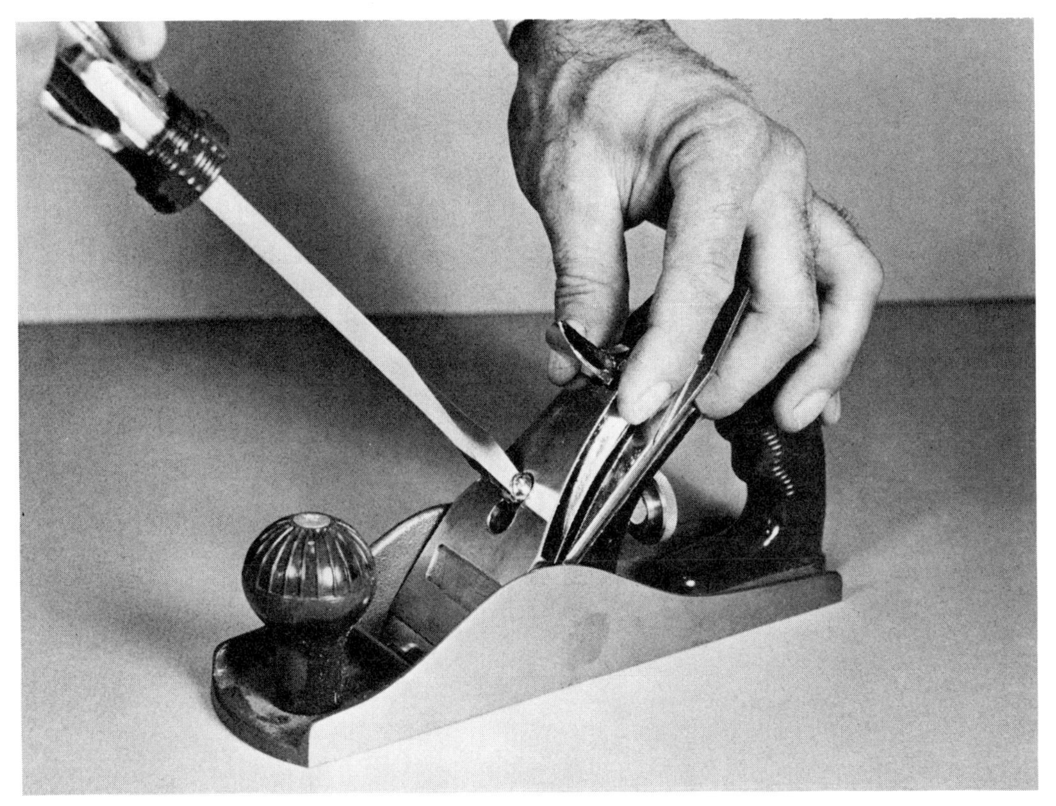

Disassemble a plane this way: Turn the lever support screw until it is loose, but not out. Then flip the top cam (rotating or sliding piece), which will remove the cap. You will have to disassemble any plane to sharpen the blade and to adjust it with the plane iron and plane iron cap.

The plane iron and plane iron cap are held together with a screw, which you can remove with the lever cap. These two components have to be separated for sharpening purposes and adjusting the depth of the cut. For general plane maintenance, give your planes an occasional steel wool buffing and a light coat of oil.

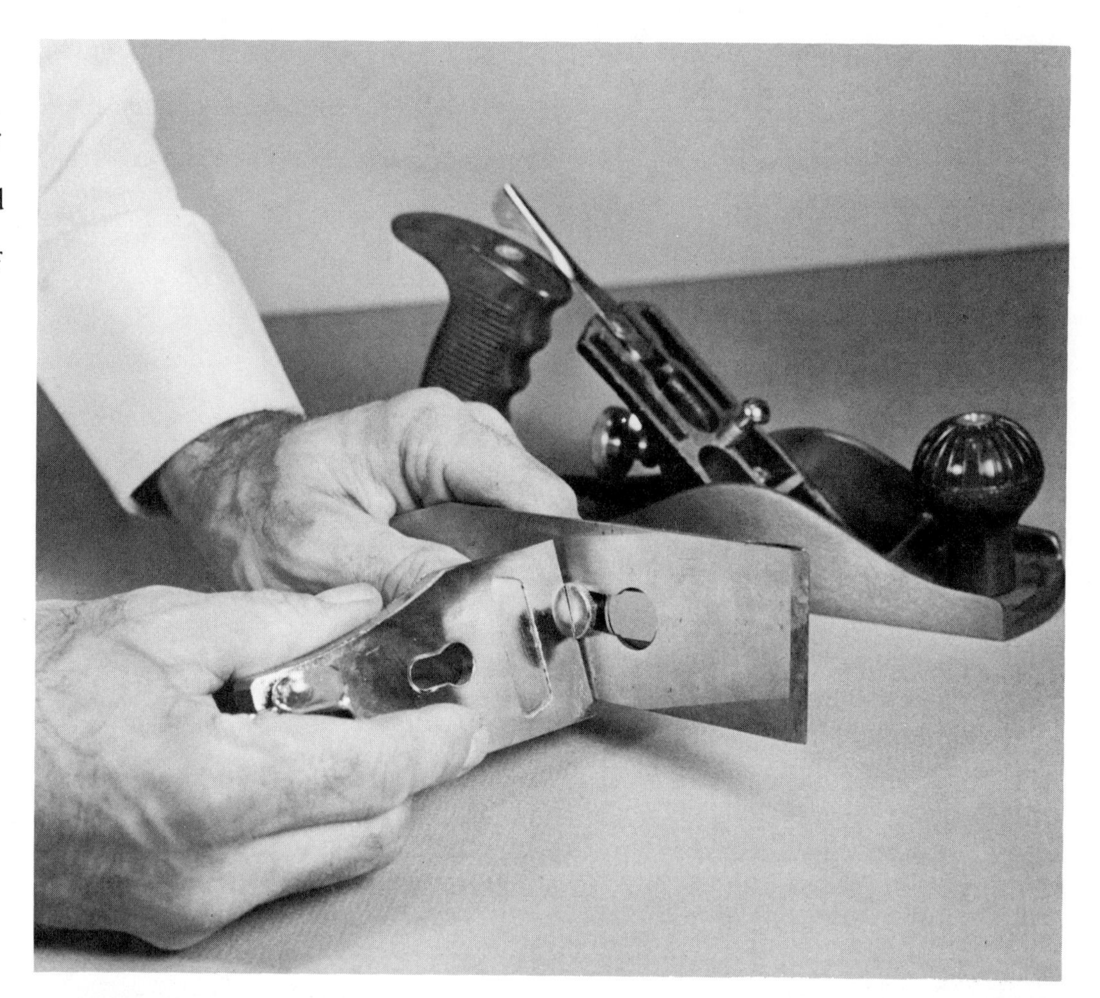

The plane iron (blade) is supported by the plane iron cap. The end of the cap should be about ¼ inch up from the bottom of the iron or blade. Adjust the cap, reassemble the plane, and test the blade on a piece of scrap. If the plane doesn't seem to cut properly, disassemble the unit and move the plane iron cap up or down.

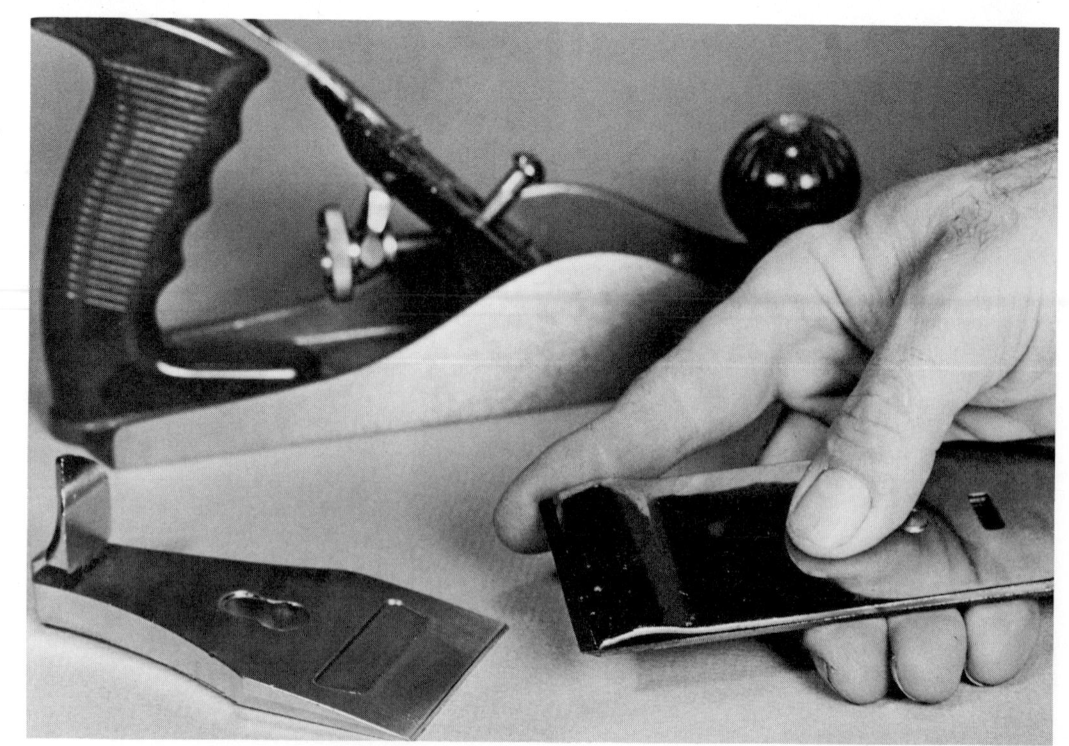

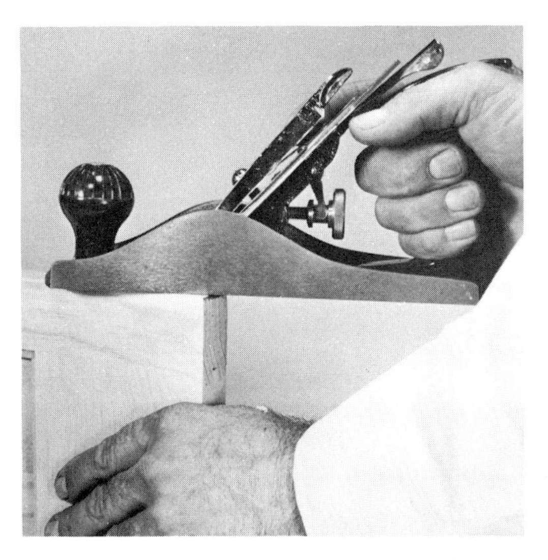

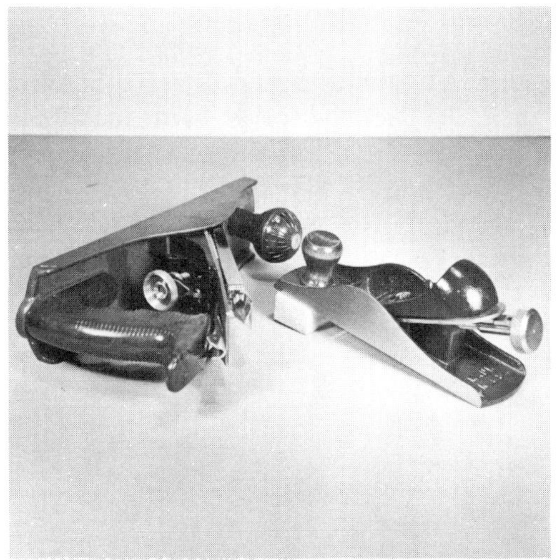

Far left:
The wood won't splinter when you're planing toward the end of the stock or across the grain, if you reverse the plane near the end of the stroke and plane in the opposite direction. Another trick: If you are planing a board with lots of knots in it, run the plane to the knot from opposite directions. This way, you won't dig into the stock or the knot.

Left:
Protect plane blades by laying the planes on their sides or blocking one end with a piece of scrap stock, as shown. When you store your planes, retract the blades to protect the sharp edges from nicks. To sharpen a blade, remove it from the cap, and run the blade along its bevel straight across an oilstone or whetstone. Maintain the bevel; then turn the blade over and stroke it straight across the stone to remove any slight burrs. It is best to start the sharpening process on the coarse side of the stone and finish on the fine side.

A power sander is ideal for most smoothing and rough sanding jobs; it will make your work easier when you have a lot of sanding to do. For your initial purchase, we recommend an orbital or reciprocal sander. Later, when you add to your equipment, you will want to buy a belt sander. This tool is tops for the heavy jobs, while the orbital or reciprocal models are best for light sanding.

Screwdriver how-to

Screwdrivers should be used for one purpose: to drive and to draw screws. Screwdrivers are not lid prys, paint paddles, hole punches, nail sets, toothpicks, or back scratchers. Standard blade and Phillips screwdrivers are workshop basics, and since they are fairly inexpensive, buy an assortment. They will be used for a host of woodworking and repair projects.

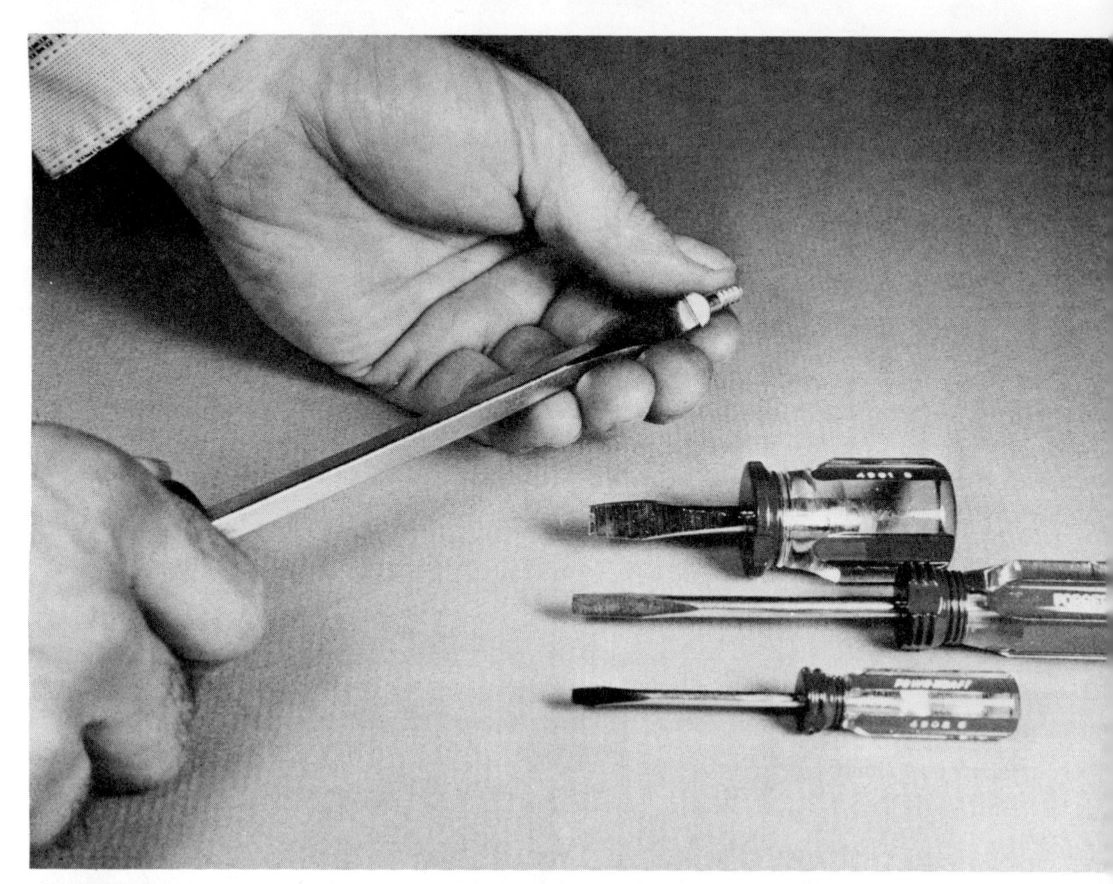

Screwdriver blades must fit the width and length of the screw slots in which they are used. If the blades are too wide, they may gouge or mar the wood around the screw hole when the screws are driven home. If the blades are too narrow, you can't get the proper turning leverage, and the driver blades may break the screw slot, or at least damage it.

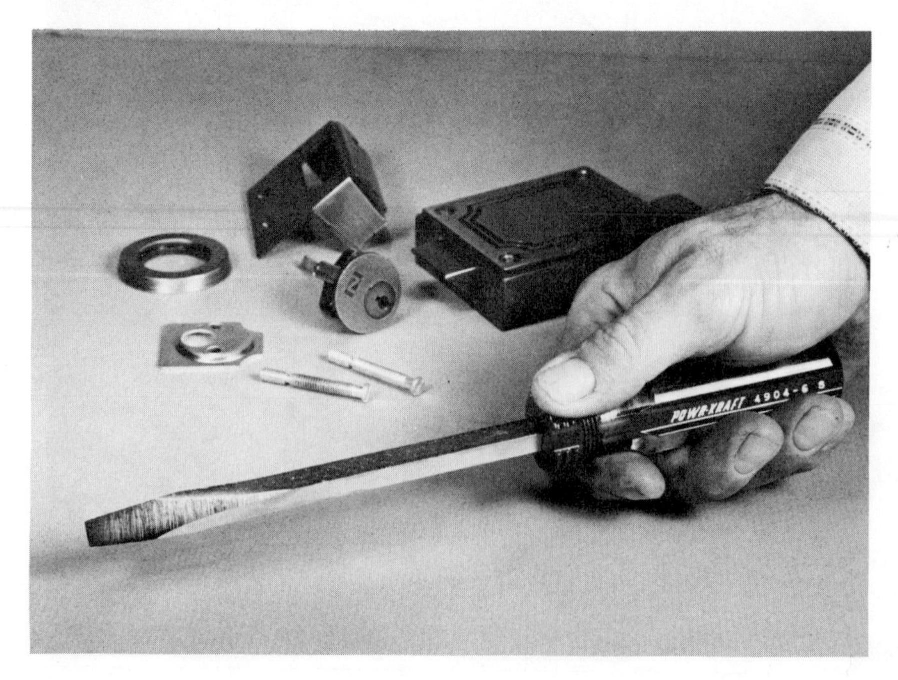

Hold a screwdriver with your fingers and the palm of your hand. Your fingers do the turning; the palm of your hand applies the necessary pressure to the screw. Always use the longest screwdriver you have to drive screws; the length gives you more leverage.

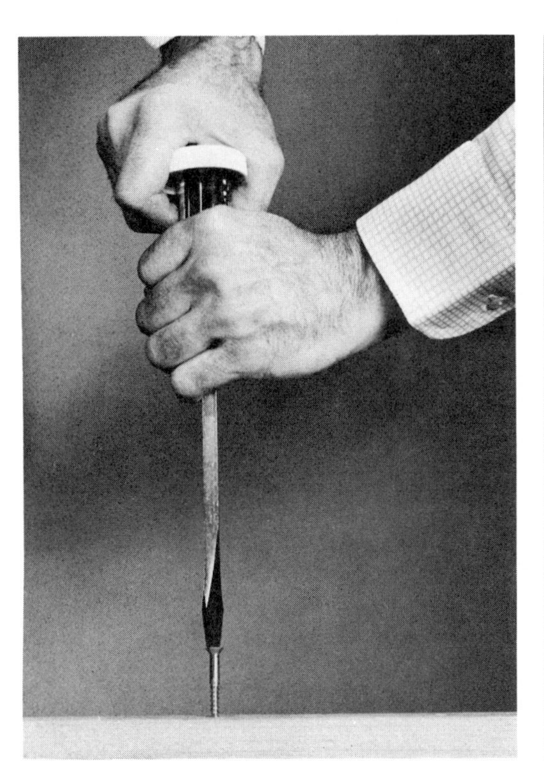

Far left:
Obtain more driving power by covering the top of the screwdriver handle with a jar lid. The lid provides more bearing surface without putting undue pressure on your hand. The other hand supplies the twisting power. If screws are really tough to drive, try coating the screw threads with soap or paste wax.

Left:
Get more twisting power by turning the screwdriver with an adjustable wrench which is locked on the screwdriver's square shank. The lid trick supplies the necessary downward pressure. As a general rule, you should always drill pilot holes for screws; use a bradawl (a small tool used to make holes in wood) to punch holes for tiny screws.

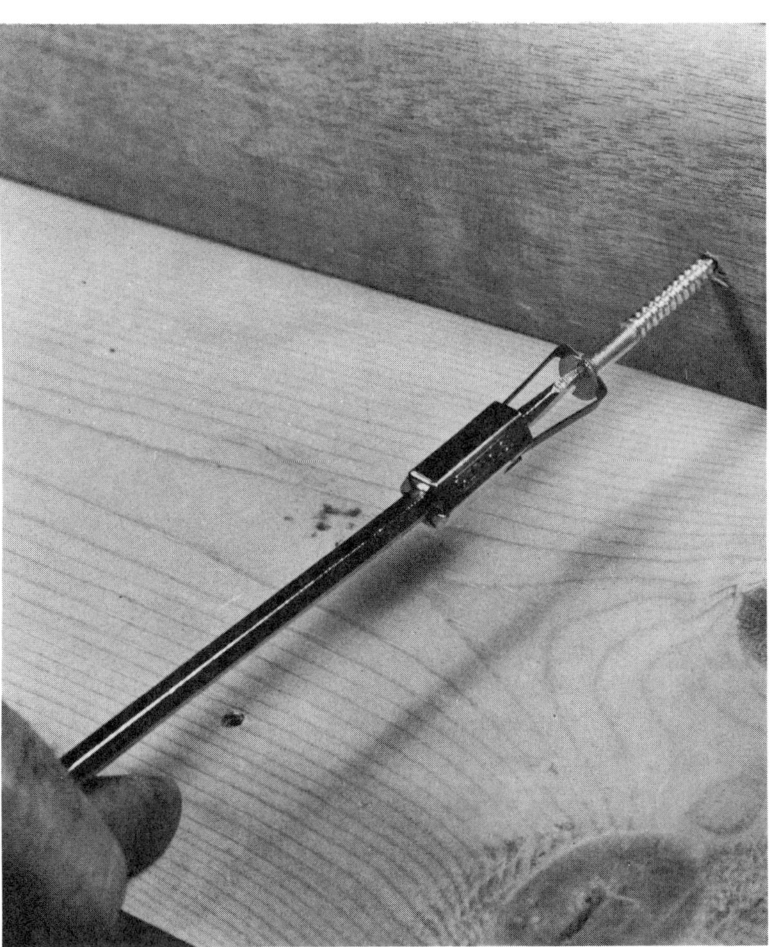

A *screw-holding outfit* holds the screw to the tip of the screwdriver blade until the screw has started into the material. The holder has a spring clamping device; once the screw has been twisted a couple of turns, jerk the holder from the screw.

If you don't have a screw holder, push the screw through a piece of masking tape to get it started into the stock. If you don't have masking tape, try a strip of newspaper or cardboard. Once the screw has started, simply pull away the masking tape, newspaper, or cardboard.

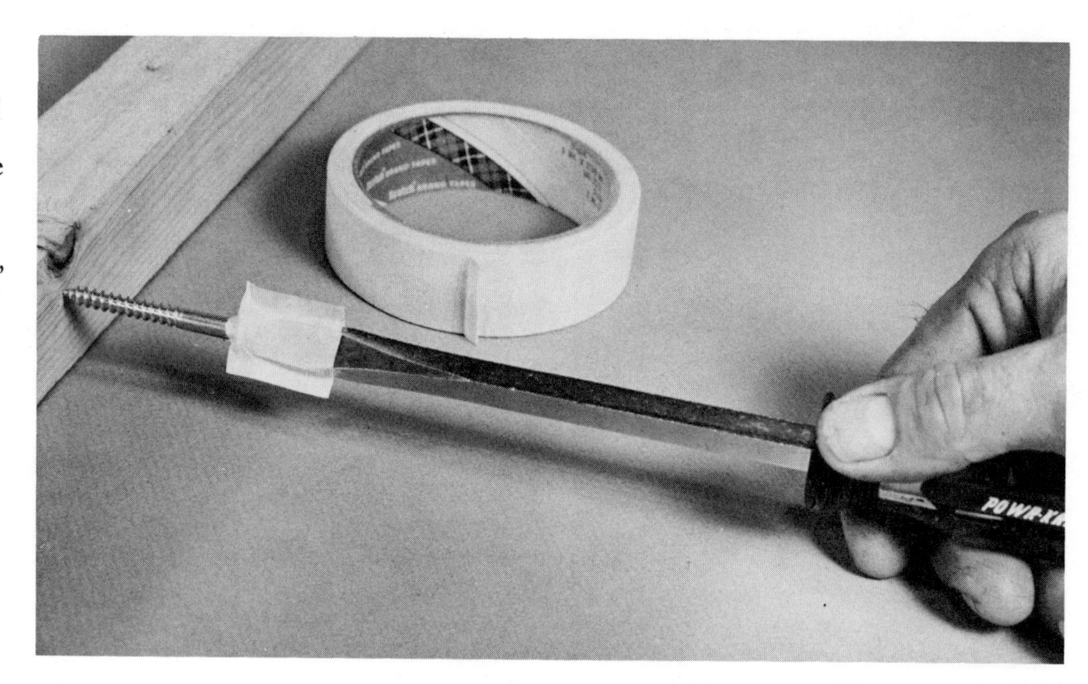

Phillips screwdrivers are for Phillips screws, which have a cross slot instead of a regular slot. These screwdrivers are manufactured in different sizes and lengths, just like the standard ones. Phillips screwdrivers are also used like their standard counterparts. If you don't have a Phillips handy, you may be innovative by using a very small standard screwdriver that fits one of the cross slots, or a common nail. You turn the nail with vise-grip or regular pliers.

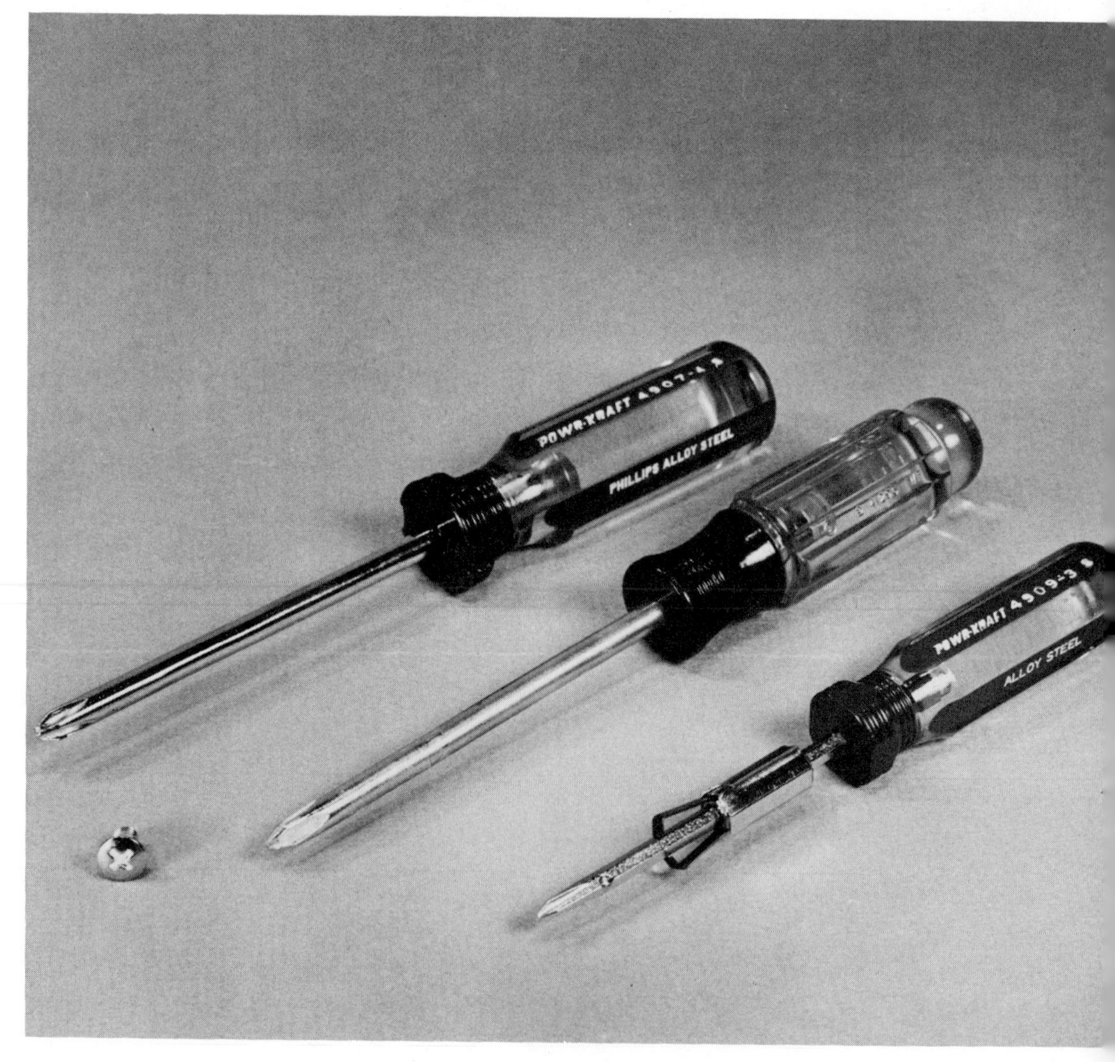

Pilot holes for screws may be drilled, punched into the material with a bradawl, or drilled with special bits like these. The shape of the bit, or drill, conforms with the shape of the screw. The pilot hole should be the same size as the threaded part of the screw. Bore or drill the hole about the depth of the screw length. If the hole will be countersunk for the screw, countersink the hole after you drill the pilot hole. If you want to counterbore the hole for the screw, drill the counterbore after the pilot hole has been bored. The depth of the counterbore should be about 1/4 inch, since it probably will be plugged with a dowel (round pin). Roundheaded screws do not have to be countersunk, but you have to predrill pilot holes for them. Ovalheaded screws usually are countersunk.

An offset screwdriver is used in an awkward spot. The driver blade is at a right angle to the shank of the driver. You can buy an offset screwdriver with a ratchet setting. This tool is flat in configuration, with the bit sticking up from the shank portion at a right angle.

A spiral ratchet screwdriver is a very handy tool when you have a lot of screws to drive. You simply insert the tip of the blade into the screw and push. Spiral ratchets also have a selection of bits you may buy, including a Phillips blade.

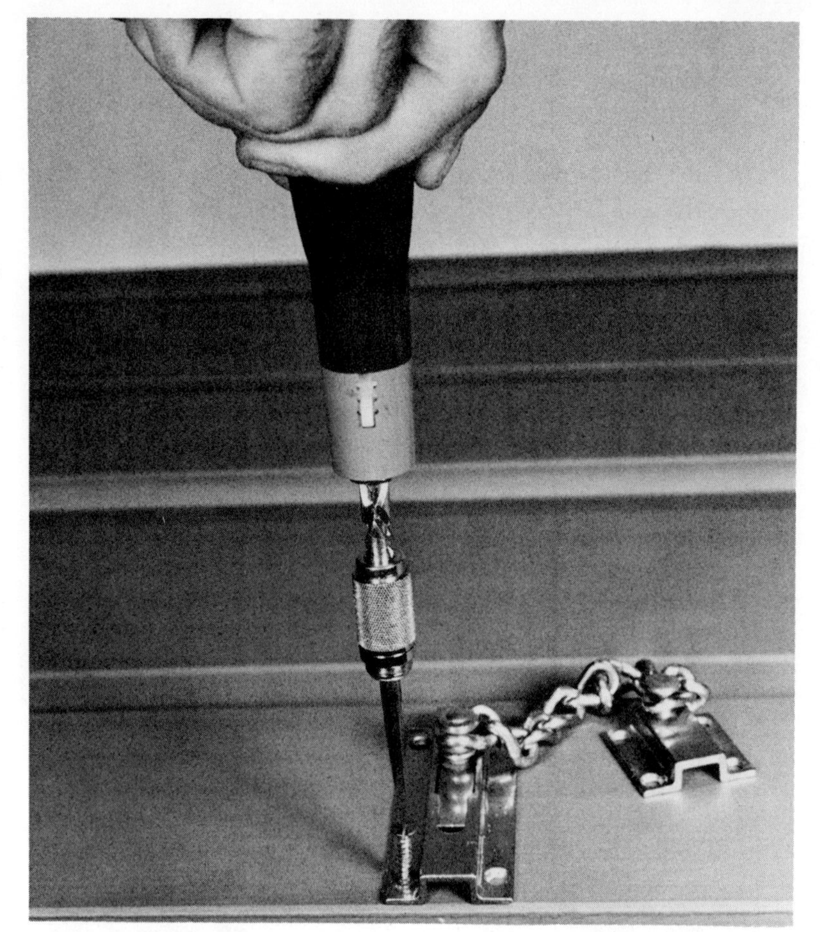

Screwdriver blades must be square; keep them this way by touching up the tips with a metal file or grinder. You can also use the file or grinder to reshape the slight taper on the shaft of the screwdriver to its tip. If the tip of the blade becomes rounded, it will tend to ride up out of the screw slot. Give your screwdrivers an occasional polishing with fine steel wool and a coat of light household oil.

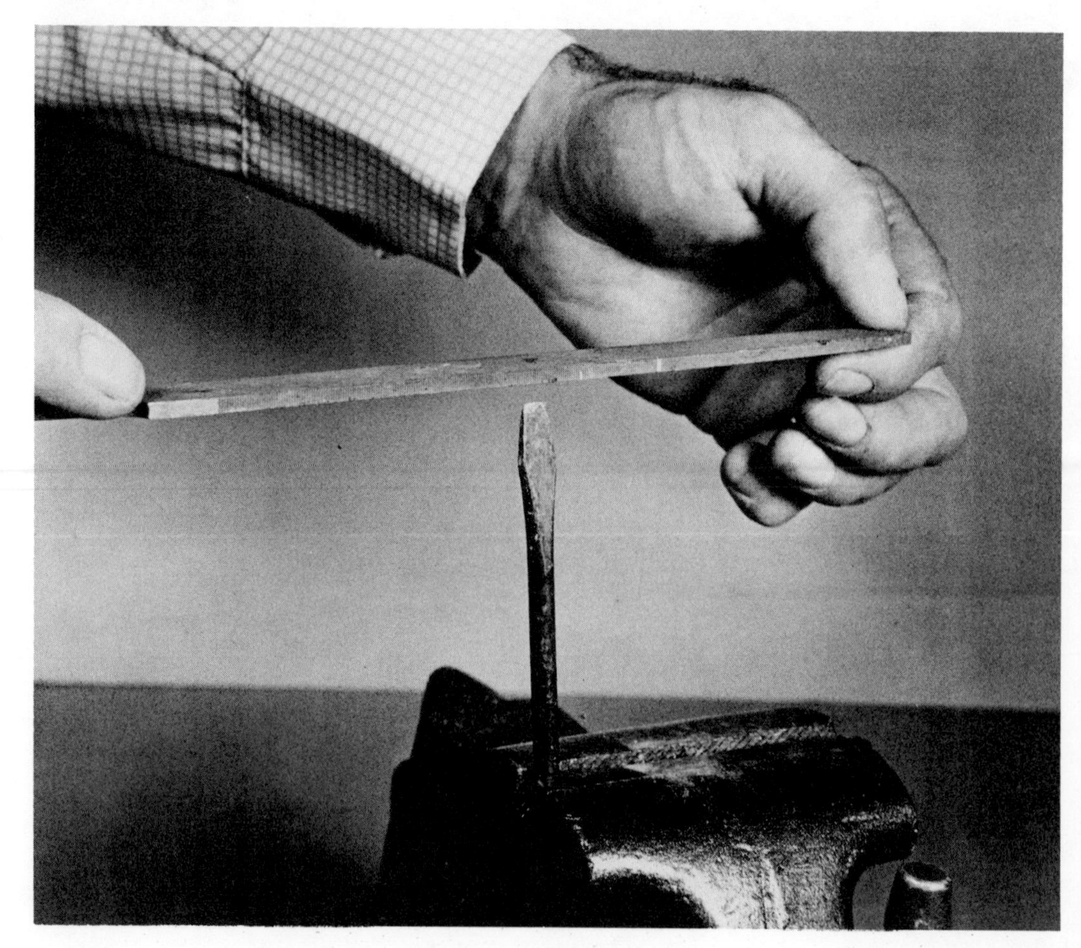

Hole boring how-to

A brace is a multipurpose hand tool and a basic one for your workshop. With a brace, you can bore holes into wood, metal, plastic, and almost any other material. Also, a brace will accept special bits such as screwdriver blades, and these will make tough projects easier.

We recommend that you buy a portable electric drill, which is a supplement to a brace and bit outfit. The power drill does not replace a brace and bit; you should have both tools in your workshop.

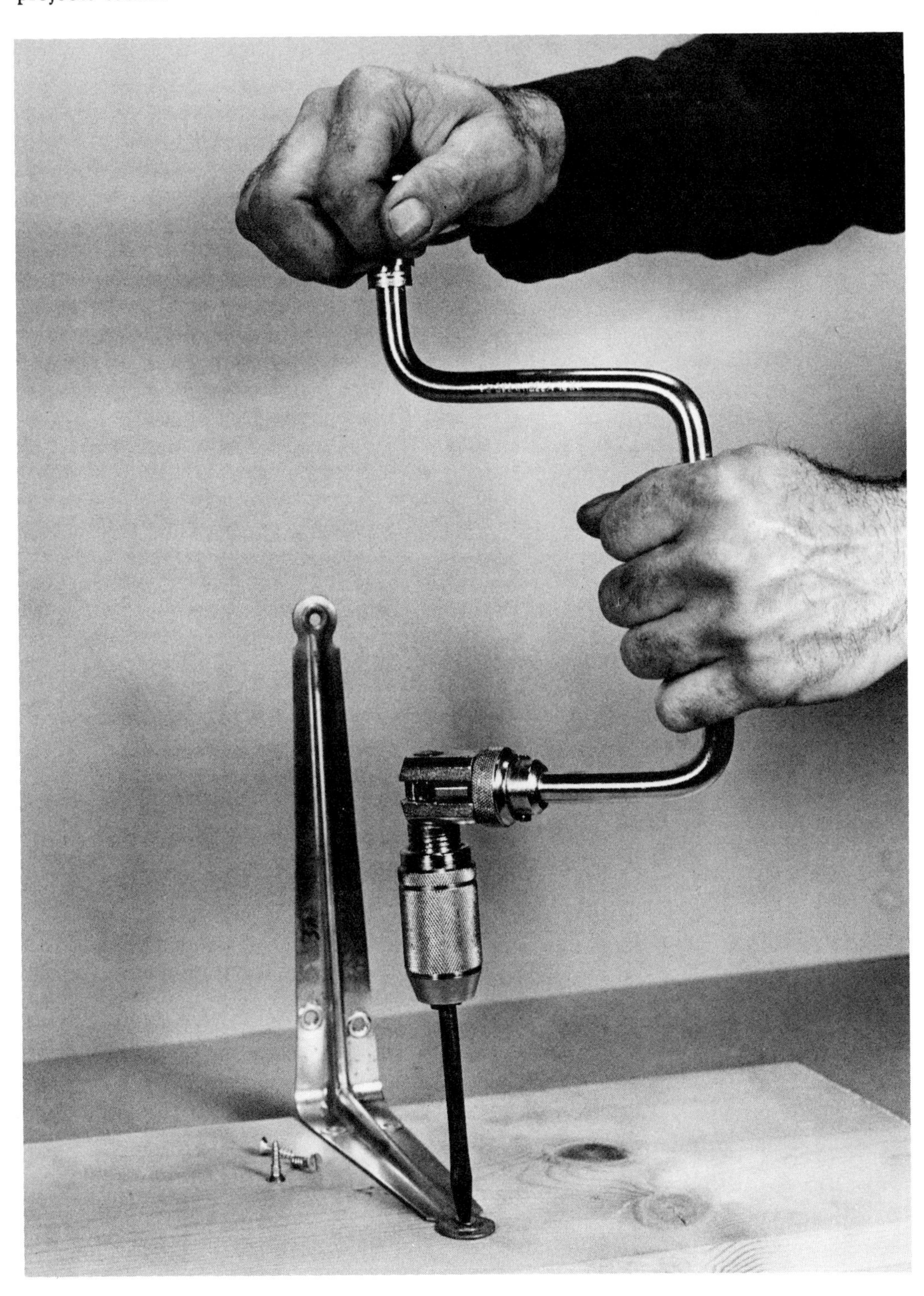

A *ratchet brace* with an adjustable chuck will handle most any project. The chuck holds bits from ¼ to 1⅛ inches in diameter; it also will handle expansive (adjustable bits) and screwdriver bits. Buy a brace with a head that operates on a ball bearing. The chuck should have an automatic centering feature, along with the ability to accept round and square shanked bits.

Auger (corkscrew-shaped) bits are sized from ¼ inch to 1⅛ inches, and they are numbered from 4 to 18. Each number graduates the size of the bit by 1/16 inch. Auger bits are from 6 to 18 inches long, although you can buy longer bits, generally called "electrician's" bits. Besides an assortment of standard bits, you'll want to invest in countersinks and screwdriver bits.

Right:
The ratchet feature is important on a brace. It lets you turn the brace in tight quarters where a hand or electric drill won't go. The sweep of the brace handle determines the force (or torque power) of the chuck. The wider the sweep, the more power you get. Handles range in size from 8 to 14 inches.

Far right:
Always mark the stock before you bore a hole. Then start the bit on the mark and hold a try or combination square against the stock and the bit. This will help you bore a "square" hole. The thrust is applied to the brace with one hand; the turning force is applied with the other. When you're boring horizontal holes, put the round handle against your body, holding it with your hand. This will provide added pressure to the bit.

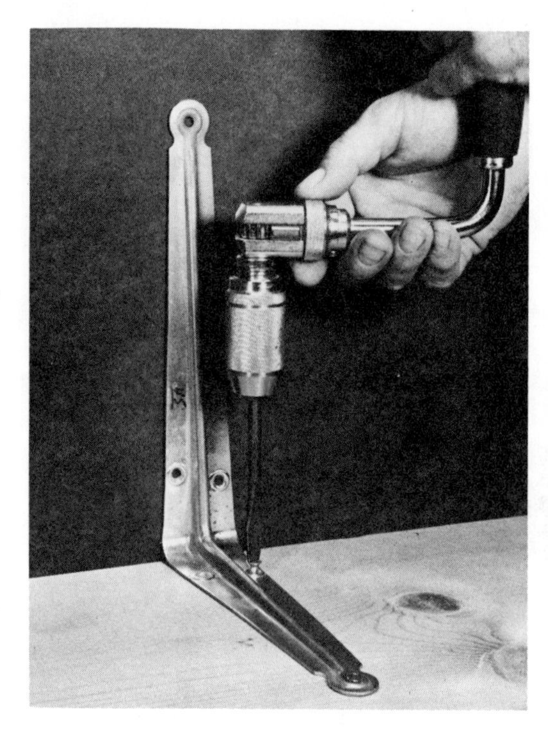

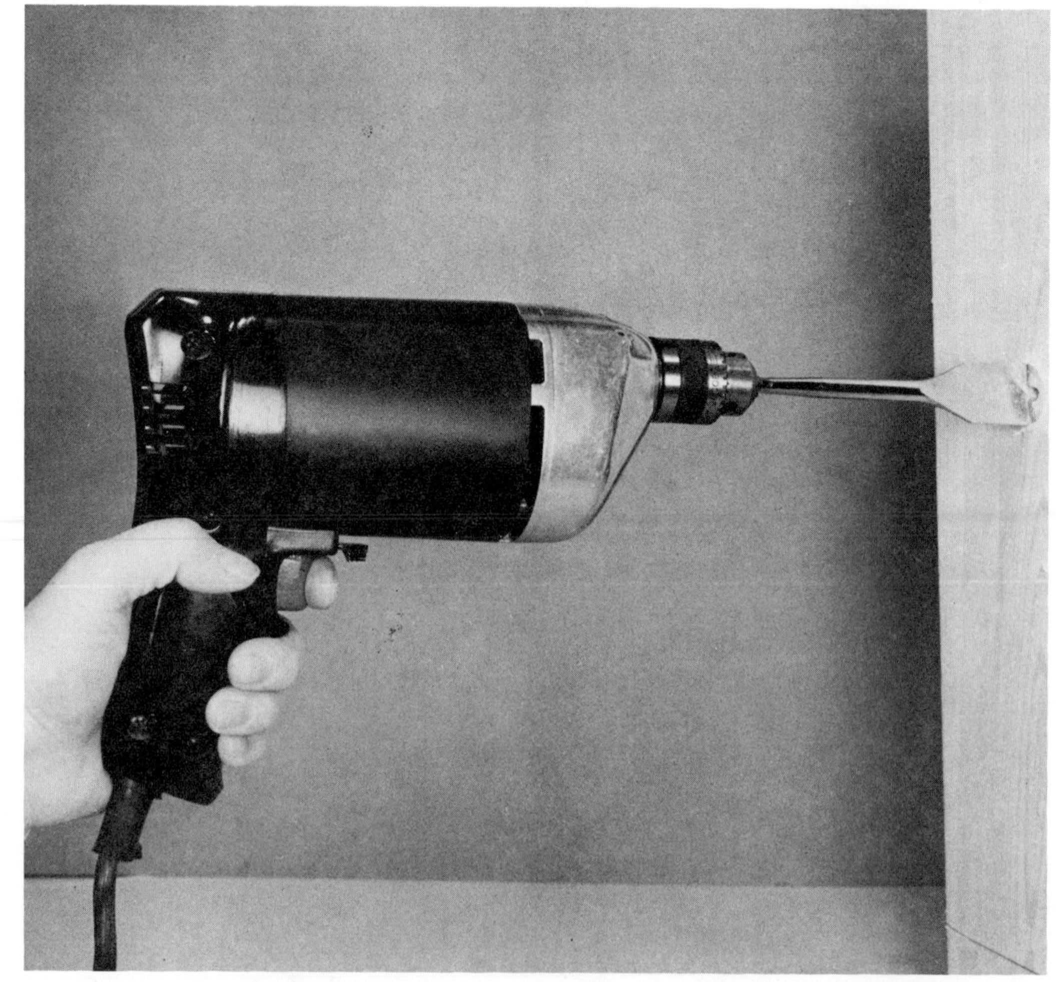

A portable electric drill, ¼ inch or ⅜ inch, is the first power drill you should buy. With different attachments, this tool may be used as a saw, a screwdriver, a buffer, and a sander. Basic accessories should include fly cutter bits with built-in pilot hole cutters, assorted twist drills, and countersink bits.

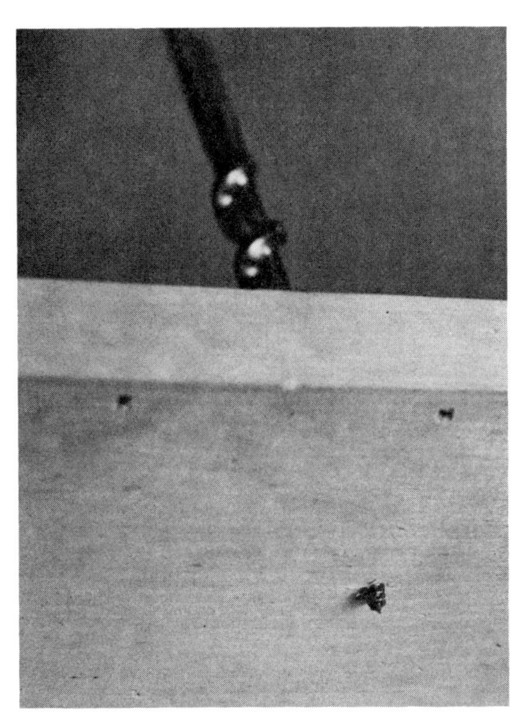

Far left:
Bore, or drill, from both sides of the stock. Bore from one side of the stock until the lead screw of the bit penetrates the other side of the stock; then remove the bit from the hole and insert the lead screw in the hole on the opposite side. This will prevent splintering and splitting the wood or other material.

Left:
Countersink holes for screws after the holes have been bored. The countersink reams (enlarges) the top of the hole so that flathead screws fit flush with the stock, or slightly below the surface. If the job calls for counterboring, do this before the initial hole has been bored. Counterbored holes are usually filled with a piece of dowel rod; drill these holes about ¼ inch deep.

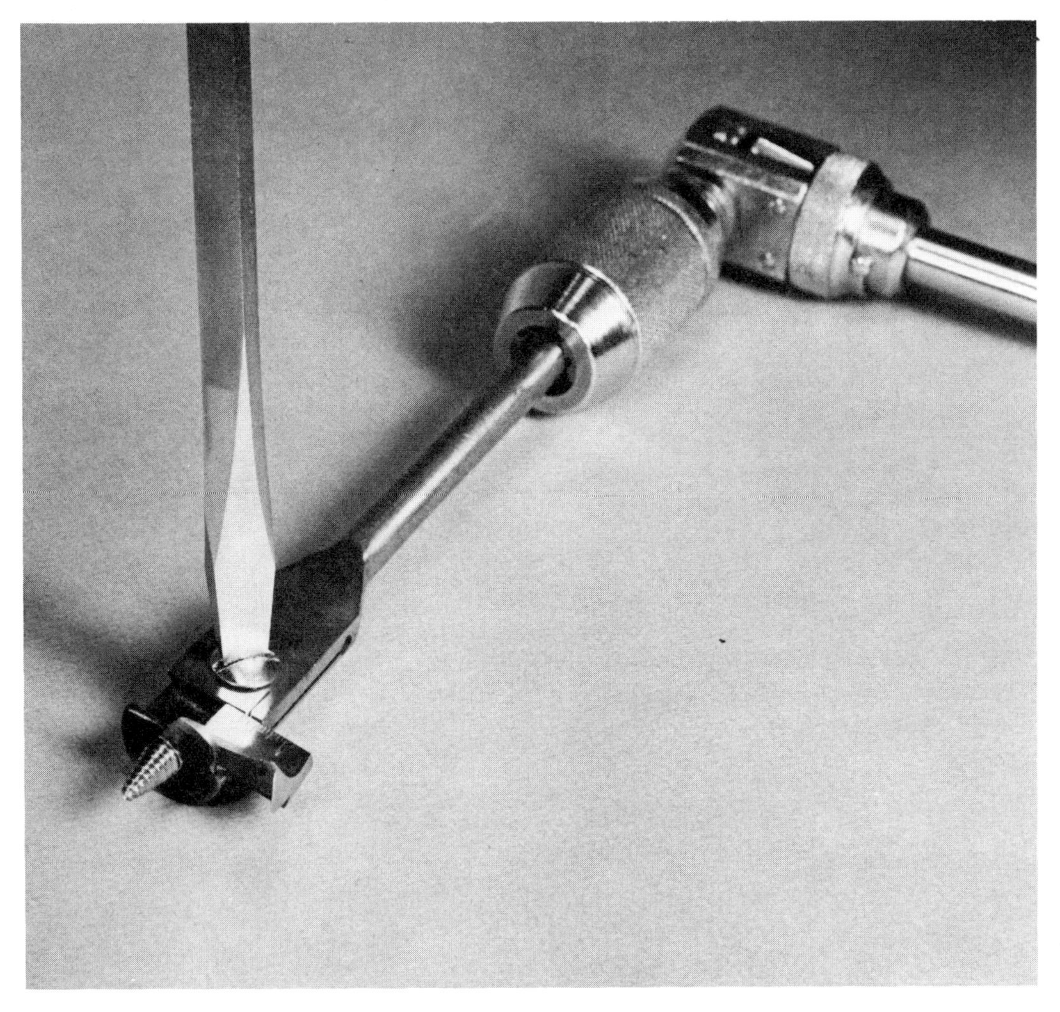

Expansive bits are adjustable for large holes. You adjust the cutter with a screwdriver; the screw locks the cutter into the position you want. Expansive bits bore holes up to 3 inches. For holes larger than this, use a keyhole saw, as explained under "Saw how-to."

A fly cutter or spade bit may be used in a brace if the shank has a square fitting. You can also buy these cutters for an electric drill. Here, the shank is round to fit into the drill chuck. Fly cutters range in size from ¼ to 1 inch, in 1/16-inch increments.

How to select bits and drills

The standard type bits and drills are the Jennings pattern (also called double twist) and the solid center bit. Other types include an expansive bit, Forstner bit, spade bit or fly cutter, countersink, and screwdriver. Also, there are masonry drills for drilling holes in concrete, concrete blocks, and cinder blocks. The charts will help you in selecting and using these tools.

Wood bits

Diameter in inches	*Length in inches*
3/8	6-1/2
1/2	6-1/2
5/8	6-1/2
3/4	6-1/2
7/8	6-1/2
1	6-1/2
1-1/8	6-1/2
1-1/4	6-1/2

Drills for masonry material

Diameter in inches	*Shank size in inches*	*Length in inches*
3/16	3/16	4
1/4	1/4	4
5/16	1/4	4
3/8	1/4	6
1/2	1/4	6
1/2	3/8	6
5/8	1/2	6
3/4	1/2	6

Drills for wood and metal

Diameter in inches	*Length in inches*
1/16	1-7/8
5/64	2
3/32	2-1/4
7/64	2-5/8
1/8	2-3/4
9/64	2-7/8
5/32	3-1/8
11/64	3-1/4
3/16	3-1/2
13/64	3-5/8
7/32	3-3/4
1/4	4
17/64	4-1/8
9/32	4-1/4
5/16	4-1/2
11/32	4-3/4
3/8	5
25/64	5-1/8
7/16	5-1/2
29/64	5-5/8
31/64	5-7/8
1/2	6

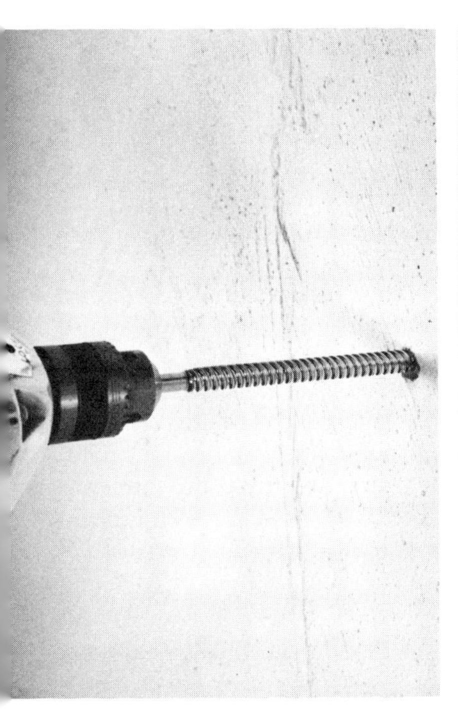

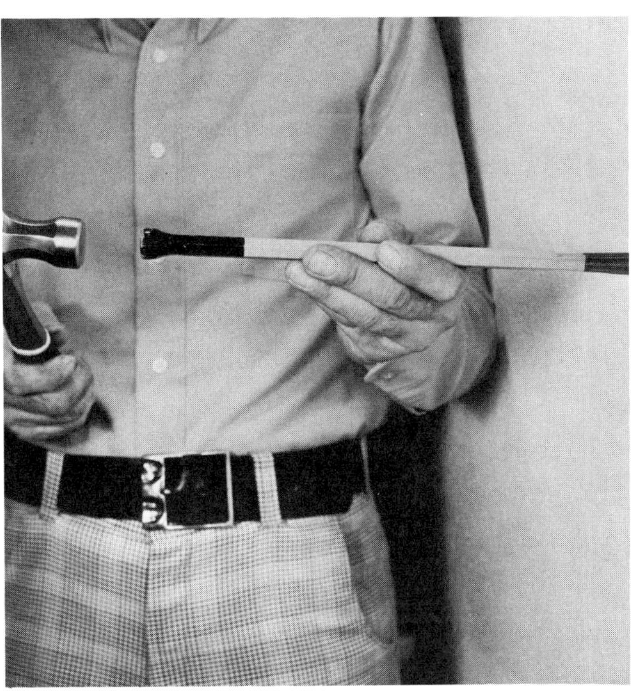

Far left:
A *masonry drill* is specially tempered to withstand the heat generated when drilling in concrete, concrete block, brick, and other masonry material. Apply medium pressure to the handle on the power drill, and remove the drill from the hole frequently. This cools the drill and removes the debris.

Left:
A *star drill* is the hand-driven counterpart of a masonry drill powered by electricity. The star drill should be struck with a sledge hammer, although a regular 16-ounce hammer may be used if the concrete material is "porous" rather than "dense." To use a star drill, twist it into the hole as you strike it. This helps to remove debris from the hole; also, the tip of the drill is fluted so that the powdered material can flow out. If you are working on a horizontal surface, remove the drill occasionally and clean out the debris with a narrow piece of aluminum sheet that has been fashioned into a "spoon."

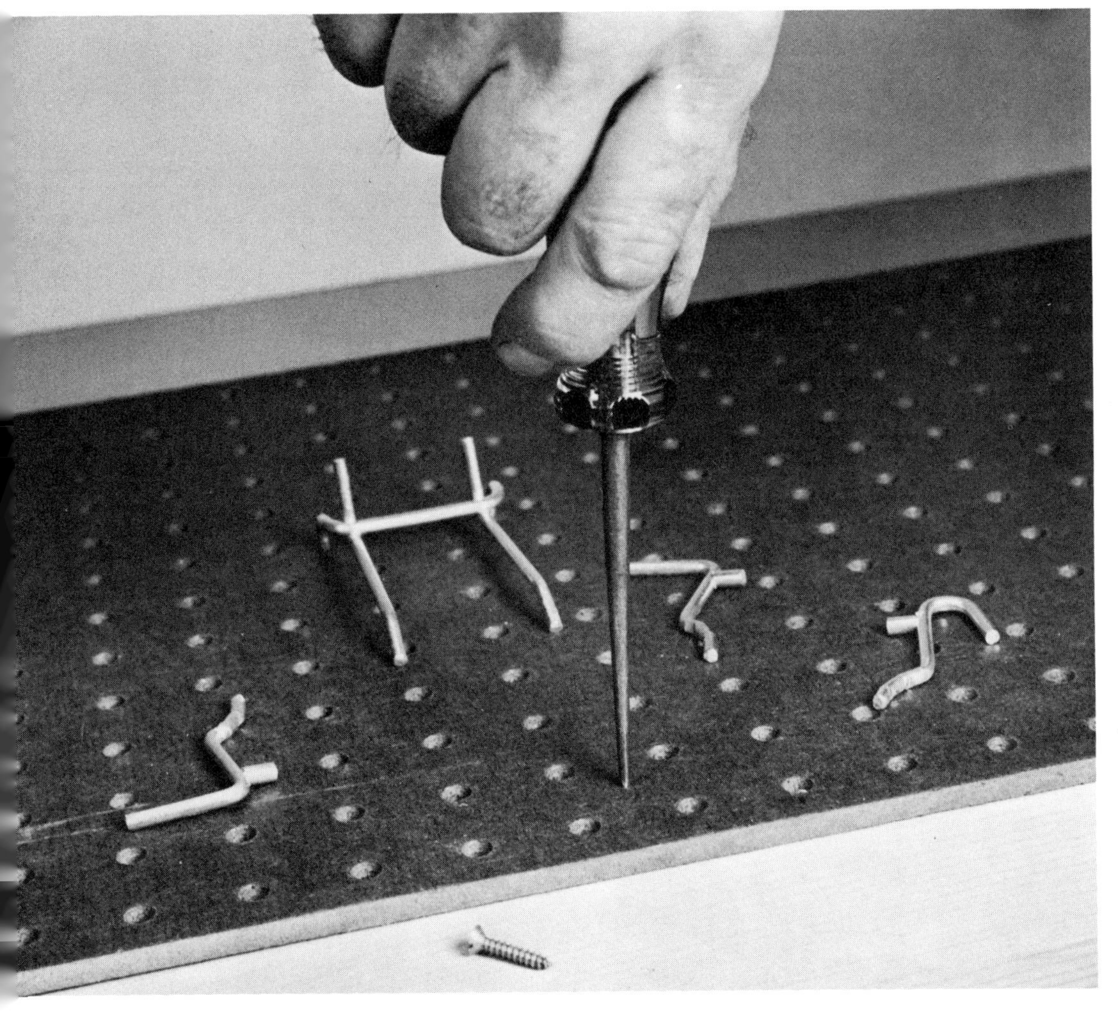

A bradawl is not a drill, but it is often used to punch pilot holes into materials for screws and nails. This tool is like an ice pick, only stronger. Keep the point of the awl super sharp by using a file or a grinder to sharpen it.

Measuring and marking how-to

A combination square may be used to square short pieces of material; it is especially useful for squaring ends of dimension lumber. A framing square is larger and has several scales on its face. The scales are used in computing angle cuts. A try square is a smaller version of the framing square; it is a handy tool for squaring ends of short or narrow materials, but not an essential one. A combination square or framing square will do the job.

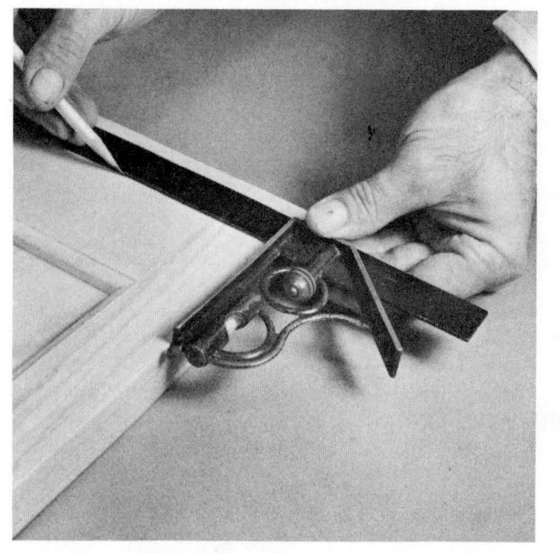

Square, level, and plumb are three magic words to remember for easy and successful carpentry. The boards and other materials that you cut must be square. Also, the framing you do has to be square, level, and plumb. A plumb is a weight suspended from a line which is used to establish a true vertical.

A combination square and a framing square are two measuring tools you must have; also, a level is needed for any remodeling or new building project.

A wise old carpenter once said: "Measure twice, cut once." To that statement might well be added "and save time and money." A miscut piece of material is costly, especially if it is clear pine or a fine hardwood such as walnut.

A flexible tape measure provides accurate measurements, and you can use this tool to compute the measurements of inside and outside curves. Quality tapes are stiff enough to be extended without support up to about 10 feet. Keep the rule lightly oiled.

Far left:
A carpenter's level gives you vertical level, or plumb (as shown), or horizontal level. When the bubble in the level is between the lines, the work is vertically true. A level is easy to use; just set it flush on the work and read the bubble.

Left:
The horizontal level is found when the bubble is between the lines. Levels are manufactured in many lengths and weights. Also, you can buy a level that will show elevation and inch rise per foot. This is a specialty item, not a basic tool; however, it is handy to have for framing a roof.

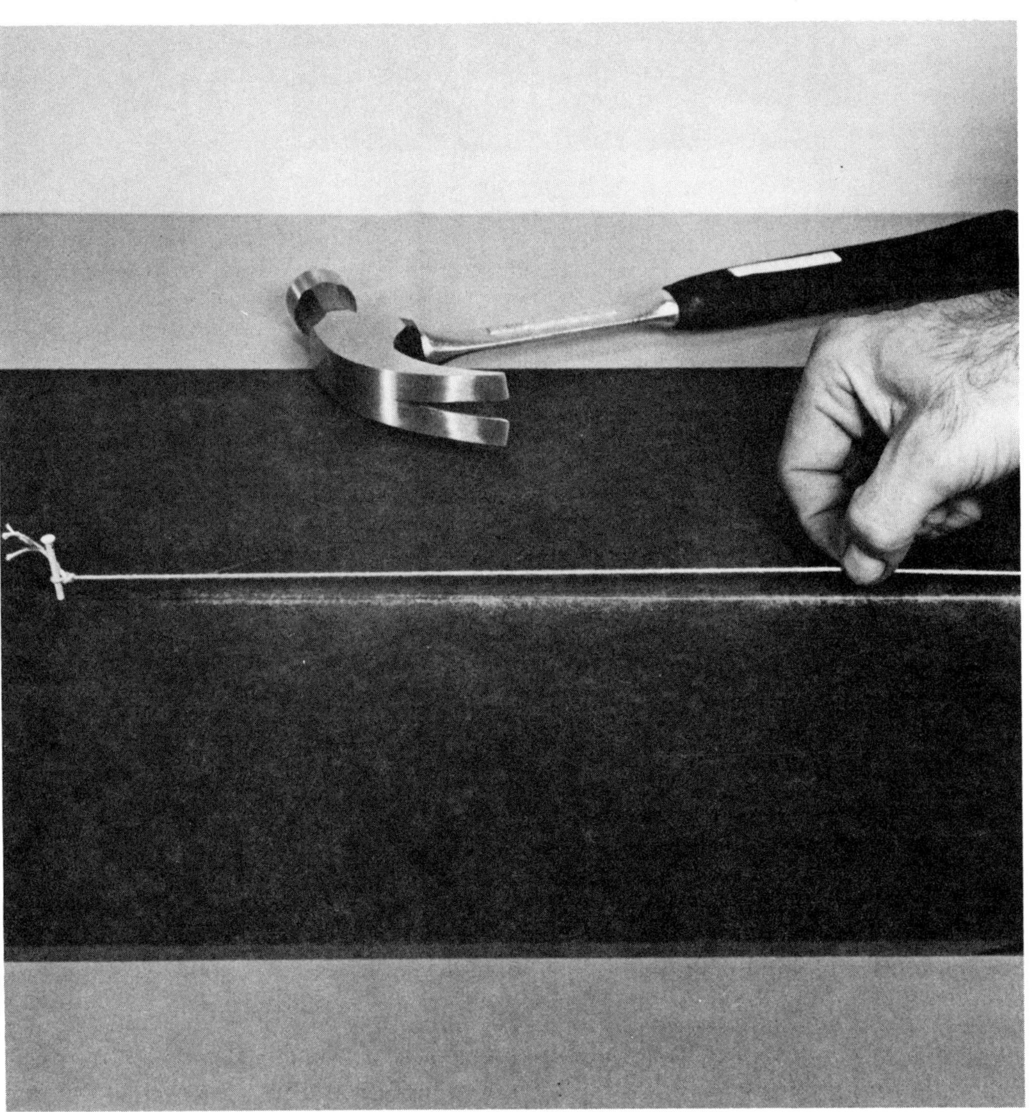

A *chalk line* is a heavy piece of twine coated with blue chalk. The line is used for marking. The line is leveled along a flat surface, pulled tight, and snapped, leaving a chalk mark on the surface. A chalk line may be used on either horizontal or vertical surfaces. You may also use a chalk line in combination with a plumb bob for true vertical lines along dimension lumber and paneling.

A *plumb bob* is a pointed weight that is fastened to a piece of twine or a chalk line. The plumb bob is usually a framing tool, designed to establish vertical (plumb) lines; you may also use it for setting pipes. Use a plumb bob in conjunction with a framing square and or level to make sure a line is true.

A *dividers or a compass* is the tool to use for scribing circles and duplicating odd shapes for trimming and cutting. For example, to cut a tile to fit around a pipe, press one leg of the divider or compass against the pipe flange (protruding rim on the shaft), and mark this configuration on the tile. The measurement to the tile is its distance to the wall. The dividers or compass has to be set to this measurement. Both tools also may be used to mark off a series of lines along a marked cutting line. Other measuring and marking tools that you may want to add to your basic workshop include a bevel gauge for duplicating angles and a caliper rule for measuring the internal and external dimensions of round material such as aluminum and steel pipe.

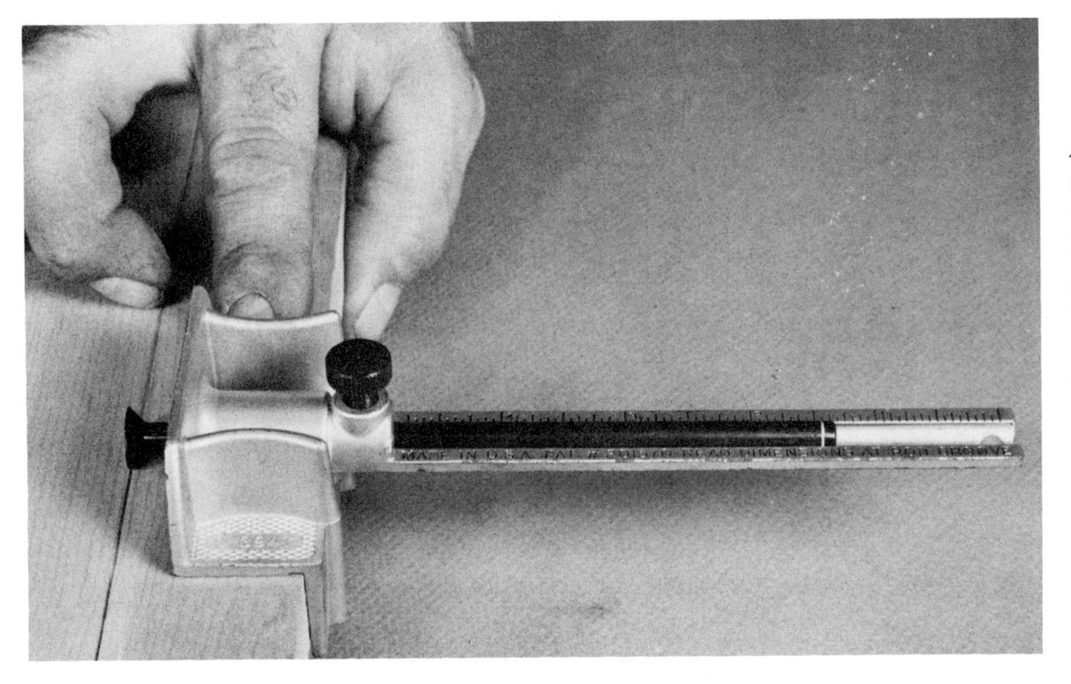

A *marking gauge* is used to mark cutting lines (with the grain) on boards that have a squared edge. Rule gradations on the gauge provide a guide for the depth of the cut, and the sliding rule is locked into position with a thumbscrew. When using the marking gauge, keep it pressed square against the edge of the stock and apply firm pressure throughout the marking process.

Pliers, wrenches, and clamps

Pliers are small pincers with powerful grips for holding, bending, or cutting, and they are invaluable for home maintenance and improvement jobs.

With pliers, you can straighten, cut, and strip wires, hold nails for driving, pull nails, and turn screws and bolts. Pliers may even be used as clamps.

Buy quality pliers which are inexpensive and well machined; the slip joints lock and unlock easily, and the jaws have deep- and smooth-cut serrations (sets of teeth).

Wrenches, like pliers, are used only occasionally in woodworking and carpentry projects, but they are basic tools you should have for home maintenance.

Wrenches are available in literally hundreds of configurations. Your first purchase should be a set of adjustable wrenches; then follow up with socket wrenches and box-end wrenches. Allen wrenches and specialty wrenches are nice to have and inexpensive, but not essential.

Clamps are basic woodworking tools. You need at least two C-clamps for clamping glued wood joints and for holding materials while they are being sawed, planed, and fastened together.

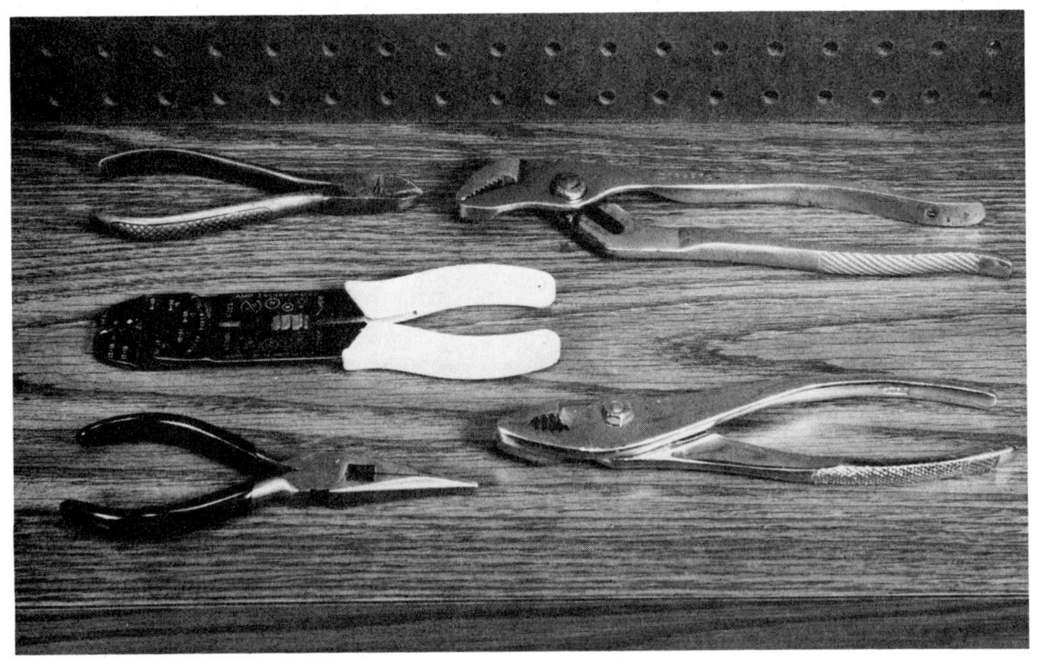

A good selection of pliers includes (from top left to right): electrician's side cutters for cutting wire, multiposition slip-joint pliers that open wide to accept large nut and bolt heads, wire strippers for stripping the insulation from electrical wire and for cutting the wire, needle-nose pliers for cutting wire and using in appliance and electrical work, and regular slipjoint pliers with two locking positions.

Vise-grip pliers have a locking device which is activated by a screw in the handle. You slip the jaws over a screw or bolt, turn the screw down until it is fairly tight, and then squeeze the handle shut. The jaws are firmly clamped on the object—like a vise. When you are using any type of pliers with serrated jaws on delicate material, such as chrome steel, pad the jaws of the pliers with adhesive bandages or cotton make-up pads. This will prevent the serrations from marring the work.

Pliers need little maintenance. If slip-joint pliers fail to lock properly, try tightening the bolt that holds the handles together. Also, give the serrations an occasional cleaning with a file card; if the serrations become worn or damaged, restore them to their original shape with a 3-cornered metal file. When working with pliers, keep your little finger under the handle; this lets you quickly open and close the pliers with a flipping action.

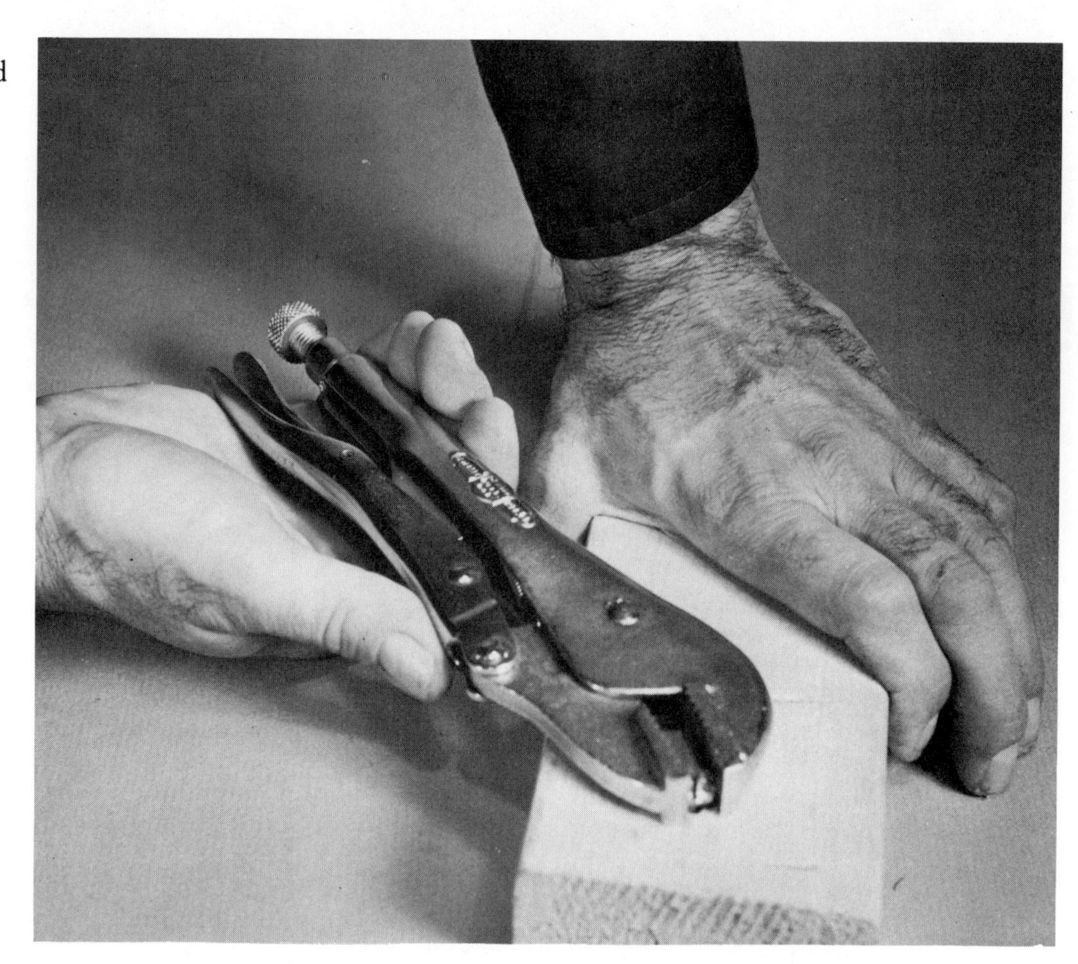

Adjustable wrenches have smooth jaws with a turn screw that operates them. The wrenches are made in a wide range of sizes and handle lengths. Always make sure the jaws of the wrench fit the bolt to be turned; if the jaws are slipping on the bolt, the wrench will strip the metal, or it may crack your knuckles. Never slip a pipe over a wrench handle for more leverage; use a larger wrench instead.

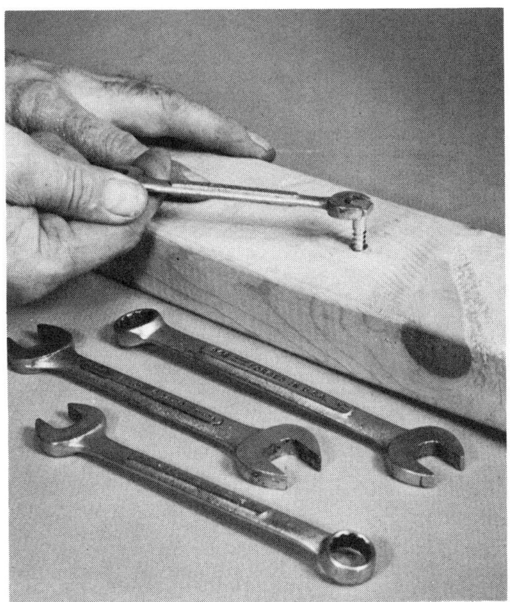

Far left:
Socket wrenches have either square or hexagonal heads to fit square or hexagonal bolts. You may want to invest in both types, since socket sets are fairly inexpensive. Handles are straight or have ratchet fittings to use in tight quarters and for leverage. Another handle feature is a push-button device that lets you snap a socket off and on with ease. Keep wrenches clean by giving them an occasional wiping with a grease solvent.

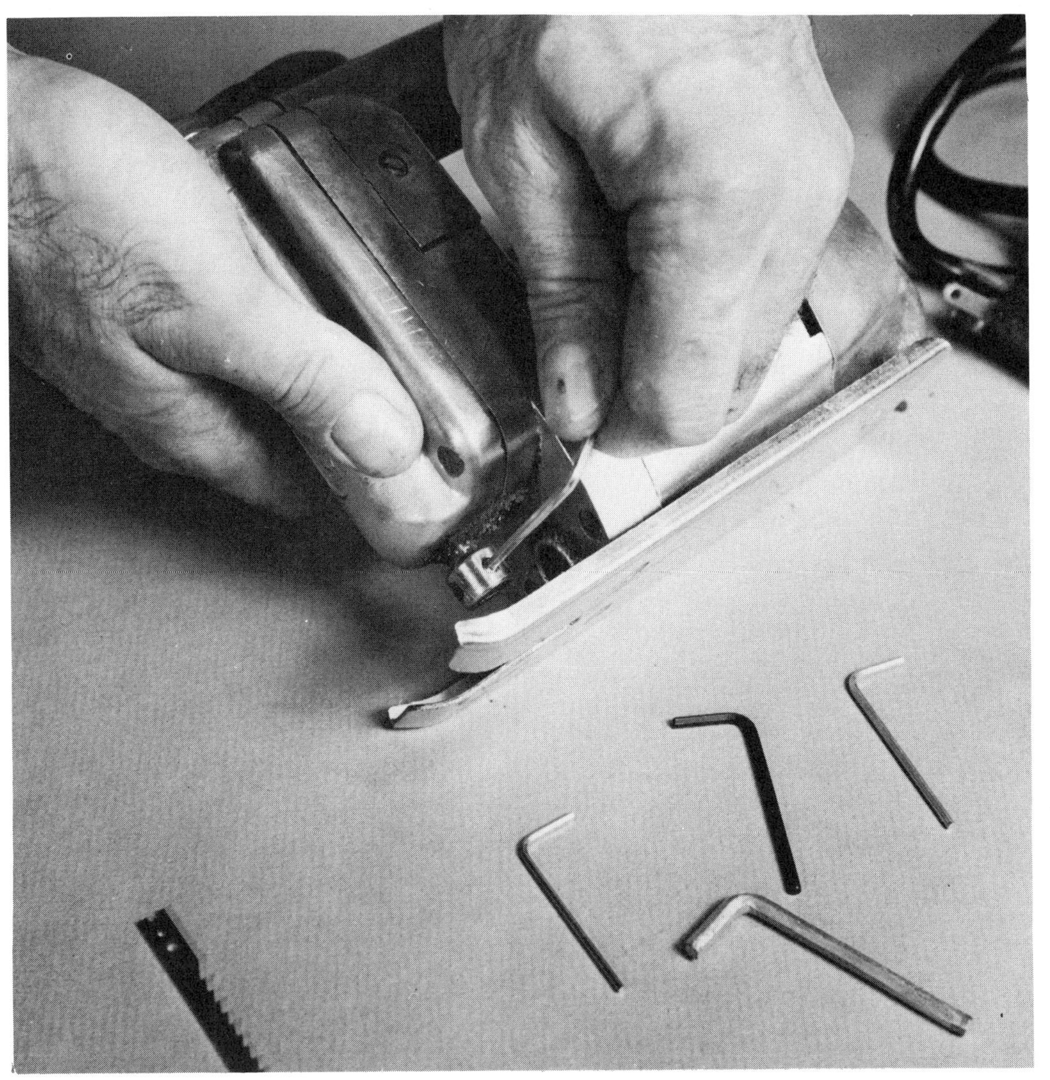

Left:
Open-end wrenches are available in sets for standard size bolts and nuts. Some have box ends on one end and open ends on the other, as shown, or, you can buy plain box-end wrenches. Nut drivers are similar to screwdrivers, with sockets that fit the nuts and bolts; they are handy for tightening a series of bolts, as in assembly work.

Allen wrenches or hexagonal wrenches fit pulleys and parts on appliances and some power tools; here, a sabre saw. These wrenches are used more for home maintenance than carpentry; the cost of a set in assorted sizes is nominal.

C-clamps are a workshop basic and can be bought in a wide range of sizes. These clamps should be clamped squarely onto the stock; the jaws of the clamp should usually be padded with a glue block (a scrap block of wood) which will distribute the clamping pressure more evenly. Slip aluminum foil between the glue block and the stock to prevent glue from sticking the block to the wood. The depth to the throat (back) of the clamp ranges from 1 to 4 inches; however, deep-throated clamps are available.

Spring clamps are lightweight clamping tools which often are used for jobs such as glueing veneer where the glue sets fast. There are several sizes available and most of them come with a vinyl pad or cushion to protect the work from being marred. More even pressure can be directed to the work area by inserting a glue block under the jaws of the clamps.

Bar and pipe clamps are basically used for clamping stock that is edge glued. The bar clamp is mounted on a flat steel bar with notches in the bar for width adjustment. The pipe clamp utilizes a length (any length) of galvanized steel pipe, and a clamp assembly holds the clamp firmly to the pipe when pressure is applied with the handle, as shown. Both models are very inexpensive.

Strap or web clamps are used to clamp irregular surfaces until the adhesive sets. Simply tighten the clamp with a wrench or screwdriver until the web is violin-string taut. The length of the web varies, up to about 15 feet. If the surfaces of the work to be clamped are slick, hold the web in position while you activate the clamp with masking tape, as shown.

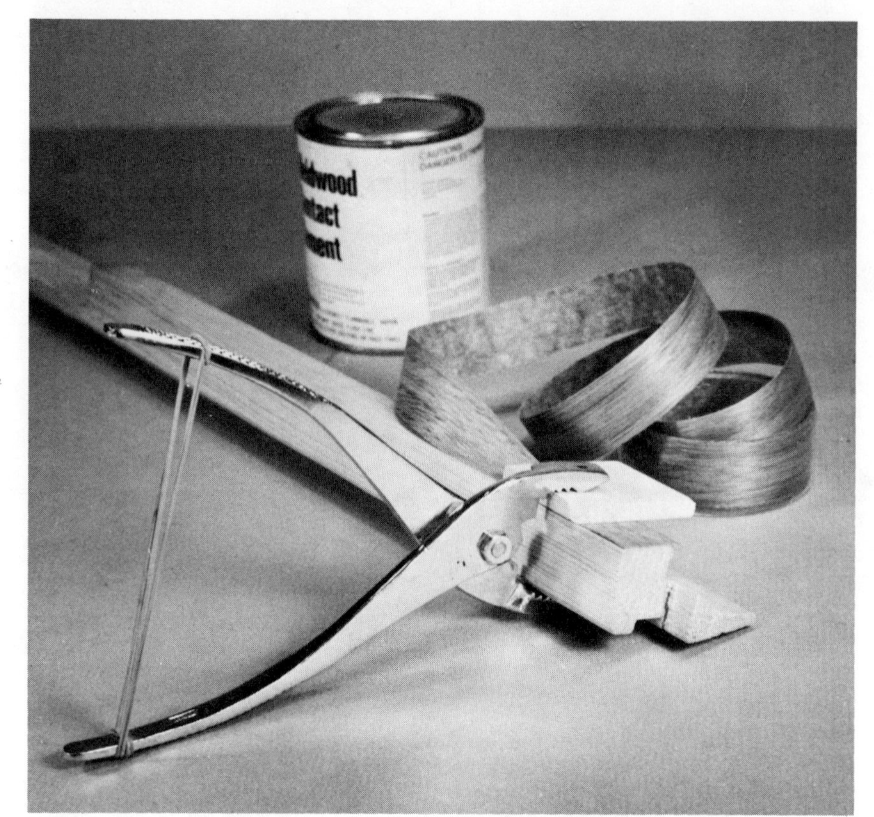

Slip-joint pliers may be used for clamping light work; a piece of twine or rubber bands firmly hold the jaws. The jaws should be padded with a glue block, adhesive bandages, or make-up pads to prevent marring the stock. You may also use vise-grip pliers or an adjustable wrench as a clamp.

A *rope clamp* is a substitute for bar clamps, since the rope can be almost any length for extra-wide projects. Wrap the rope around the project being glued and tighten it with a piece of scrap wood, as shown. The scrap may be held with a nail when the tension on the clamp is accurate. Other woodworking clamps include a miter clamp for holding mitered joints, and an edge and a hold-down clamp for holding edge glue jobs. Edge clamps are especially designed for work that is too long or wide for a C-clamp or bar clamp. Hand screws are made of wood and very expensive, but they are very practical for cabinet work since the clamps may be set at almost any angle. Sizes range from 6 to 14 inches, and the jaws open from a minimum of 3 to 10 inches.

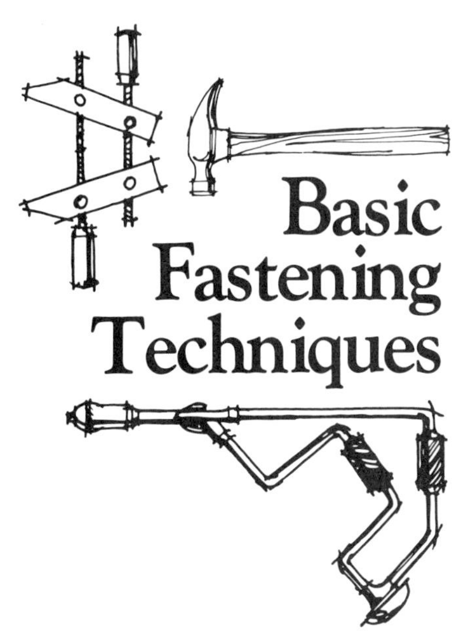

Basic Fastening Techniques

Nails, screws, bolts, and adhesives are the fasteners used in carpentry. Joinery techniques also are included, because the method used to fasten components together often depends on the joint between two pieces of material. Some joints are stronger and some are more difficult to construct than others.

We've attempted to keep fastening and joinery techniques very basic in *Easy Home Carpentry* for two reasons:

1. There are only several joints that may be cut with basic hand tools.
2. As you progress in woodworking and joinery, you will automatically be able to cut more difficult joints and use more sophisticated fasteners. You will have both the skills and the equipment.

Common nails

Nail sizes are indicated by inches and "penny" sizes. The penny size is designated by the letter "d"; for example, a 5-penny nail is written as 5d. The penny size also refers to length; for example, a 5d nail is 1¾ inches long. The nail diameter almost always increases with its length; but some nails used for special jobs, such as floor brads, come in only one size. And some nails, such as shingle nails, come in only one diameter, but in several lengths.

d (or "penny") size	*Length in inches*	*Gauge*
2	1	15
3	1-1/4	14
4	1-1/2	12-1/2
5	1-3/4	12-1/2
6	2	11-1/2
7	2-1/4	11-1/2
8	2-1/2	10-1/4
9	2-3/4	10-1/4
10	3	9
12	3-1/4	9
16	3-1/2	8
20	4	6
30	4-1/4	5
40	5	4
50	5-1/2	3
60	6	2

Nail types include (from left) a galvanized finishing nail, box nail, machine nail (used in a power-driven hammer), ringed-shank nail, and a common nail. These nails, with the exception of the machine nail which is shown for comparison, are the types you'll use for most of your carpentry projects. Other nails include flooring brads, shingle nails, nails for asphalt roofing, sheathing nails, lath nails, gypsum wallboard nails, tacks, and brads.

Finishing nail sizes

Nail size depends on the thickness of the wood you are fastening. The nail should be 3 times longer than the thickness of the material you are driving the nail through. Thus, two-thirds of the nail will be in the material that you are putting the nail into. Always try to fasten a thin piece of material to a thick piece of material. Hence, the "length of nail" rule. Nails hold better when they are driven into the wood at an angle; strive for this.

Concrete nails are made with round, square, and fluted shanks; you may buy them in lengths from ½ inch to 3¾ inches.

Specialty nails include aluminum, copper, brass, stainless steel, bronze, and monel metals. These nails are used when corrosion or rusting is a problem. Match the metals when you use these nails in metal: aluminum nails in aluminum and copper nails in copper.

Common nails of galvanized steel are used in framing and rough carpentry projects. Finishing nails, also galvanized, are used for trim and cabinet projects, and they are usually countersunk with a nail set. Casing nails are similar to finishing nails, but heavier in body. Ring- or ringed-shank nails are used in framing, flooring, and for general projects when extra fastening strength is necessary. Other nails are classified as to use, for example: "roofing nails," "upholstery nails," "gypsum board nails."

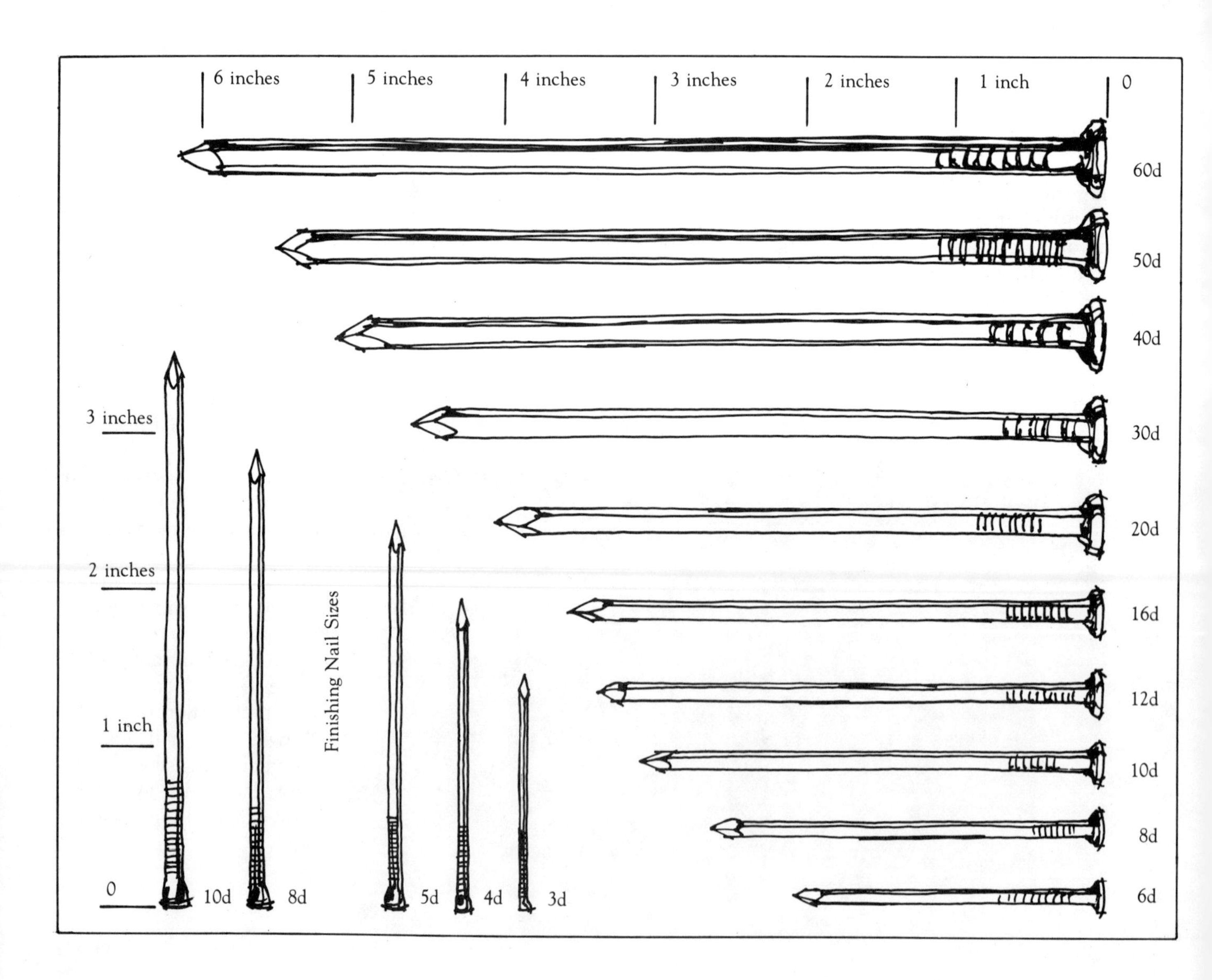

Common screws

Screw sizes are numbered according to length and gauge (thickness). The length is noted in inches, for example: A No. 16 wood screw is 2½ inches long. Lengths range from about ¼ inch to 6 inches, and the sizes usually are approximate.

In buying screws, use the same length formula as for nails: One-third of the screw length should go through the top piece of material; two-thirds of the length should go into the base material.

You can buy washers for screws to match the head type; they are classed as "flat," "flush," and "countersunk." Washers provide more bearing surface for the screw heads, and they help to prevent marring the wood when removal is necessary.

Here is a list of common screw sizes:

No. 2	¼ to ½ inch
No. 3	¼ to ⅝ inch
No. 4	⅜ to ¾ inch
No. 5	⅜ to ¾ inch
No. 6	⅜ to 1½ inches
No. 7	⅜ to 1½ inches
No. 8	½ to 2 inches
No. 9	⅝ to 2¼ inches
No. 10	⅝ to 2¼ inches
No. 12	⅞ to 2½ inches
No. 14	1 to 2¾ inches
No. 16	1¼ to 3 inches
No. 18	1½ to 4 inches
No. 20	1¾ to 4 inches
No. 24	3½ to 4 inches

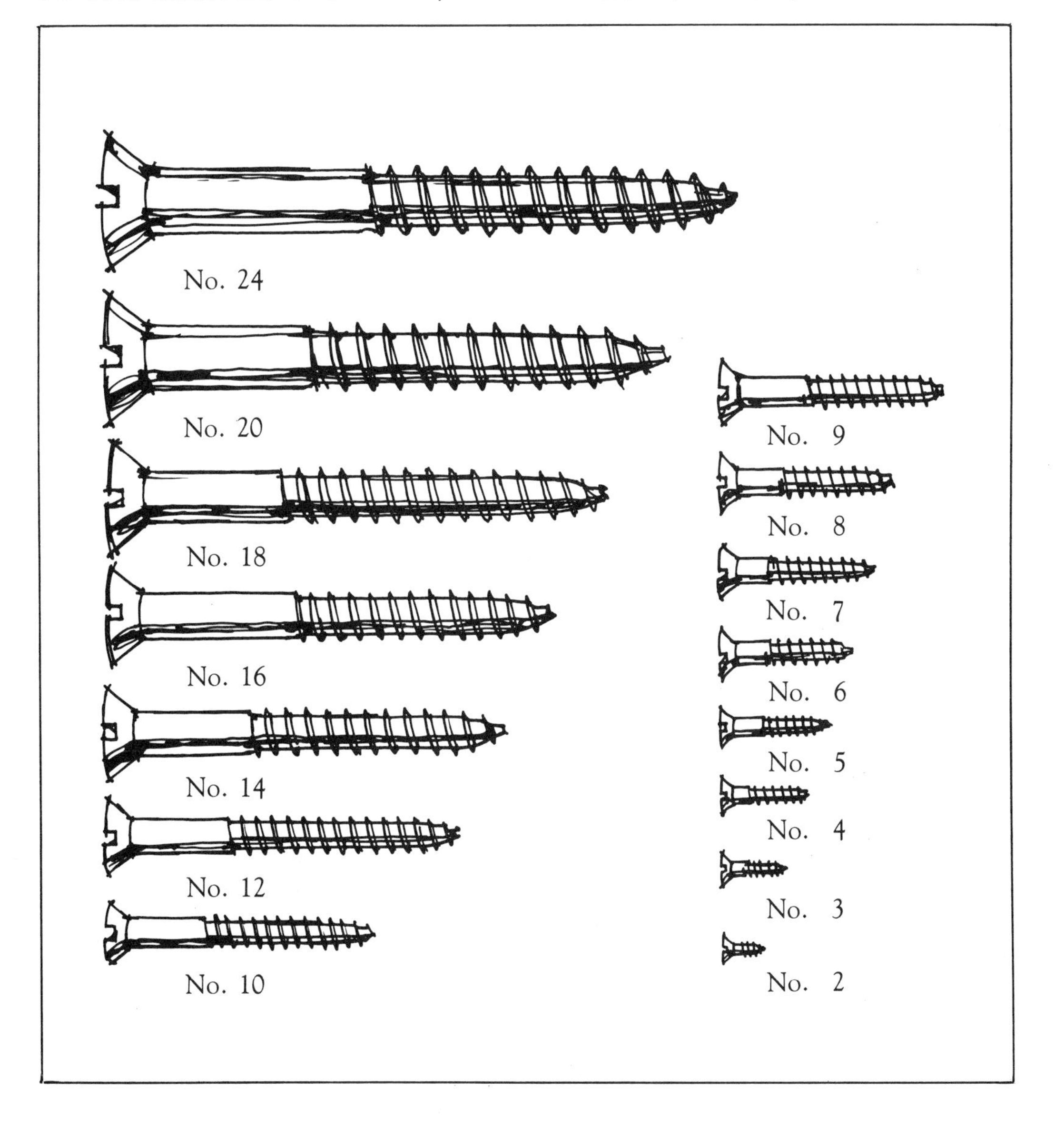

Screw types include (from left) oval head and flathead screws that are countersunk, and roundhead screws that usually are not countersunk. There are 2 basic slot types: plain slotted and Phillips slotted. Also, a stopped slot screw is available. Screws are made of steel, brass, bronze, and aluminum. Specialty screws are coated with a plating of brass, zinc, chrome, or cadmium.

Specialty fasteners you'll use often

Right:
Molly bolts are used for fastening objects to hollow core construction such as plaster and gypsum wallboard. To use Molly bolts, which are manufactured in many different lengths and diameters, first drill a hole in the material. On the packages of bolts you will find the drill size as well as the bolt size needed for a given wall thickness. Insert the bolt into the hole to make sure it fits snugly. Then activate the back of the bolt by screwing it down.

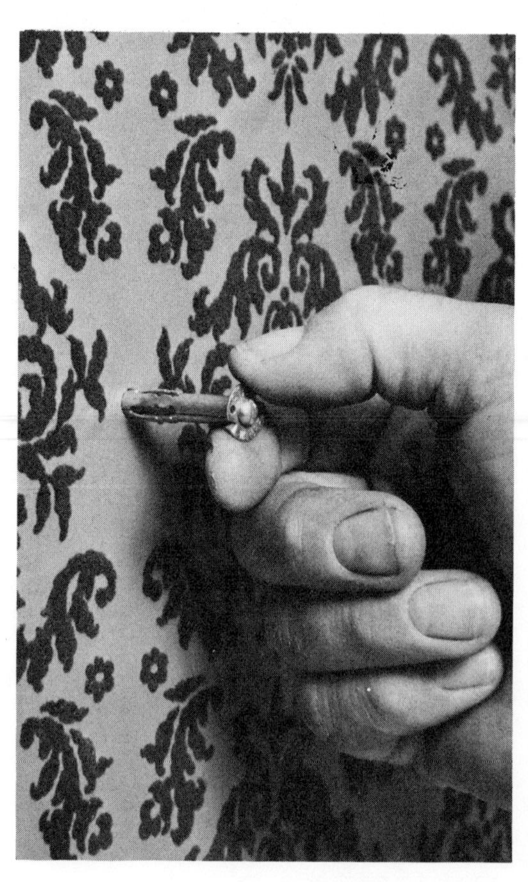

Far right:
A flange spreads in back of the wall when the bolt is screwed down on the Molly fastener. The flange grips the wall, holding the fastener firmly in place. To hang the object, remove the bolt (fastener will stay in place) and run it through the object to be secured to the surface.

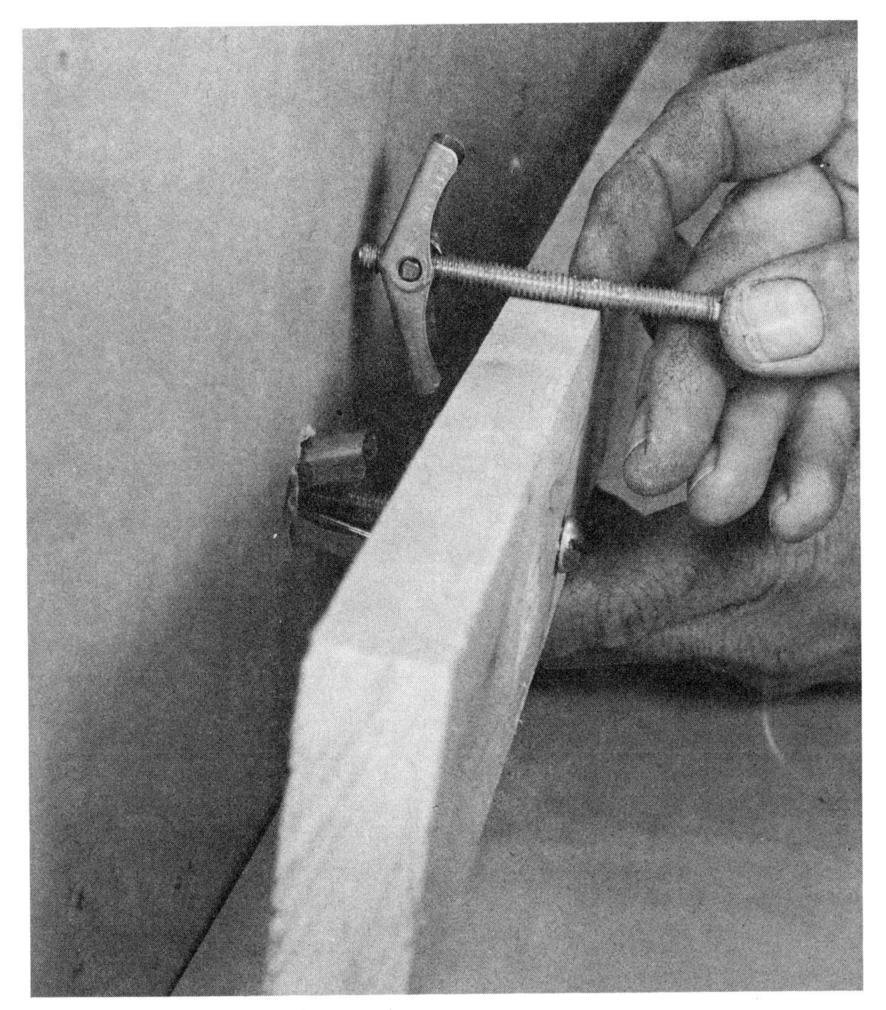

Toggle bolts are for use in hollow core construction. You drill a hole through the object to be supported and through the wall between the studs (or in a hollow of a concrete block). Insert the bolt and its flange into the hole and through the material. The flange is spring loaded so that it compresses.

The flanges of toggle bolts snap open on the other side of the opening. You then screw the bolt until the object to be supported is firmly mounted. If you ever want to take the object off the wall, remove the bolt, and the flange will drop into the hollow of the wall or block.

Lag screws or bolts are large screws with flat, square heads. Lag screws have to be driven with a wrench. These screws have great holding power; they generally are used to fasten 2-inch (or thicker) dimension lumber and timbers. Lengths run from 1 to 16 inches and diameters from ¼ to 1 inch. Always drill pilot holes for lag screws.

Masonry anchors are for fastening sills, plates, and other materials to concrete, concrete block, and brick. Large lead anchors are used with lag screws; smaller size anchors are used with regular wood screws. As you drive the screw into the anchor, which has been inserted into a hole in the masonry surface, the anchor expands, gripping the side of the hole. Drill the hole the same depth as the length of the anchor. The top of the anchor fits flush with the masonry surface. Plastic and fiber anchors are also made for light screws and materials; these lightweight anchors are sometimes called "plugs."

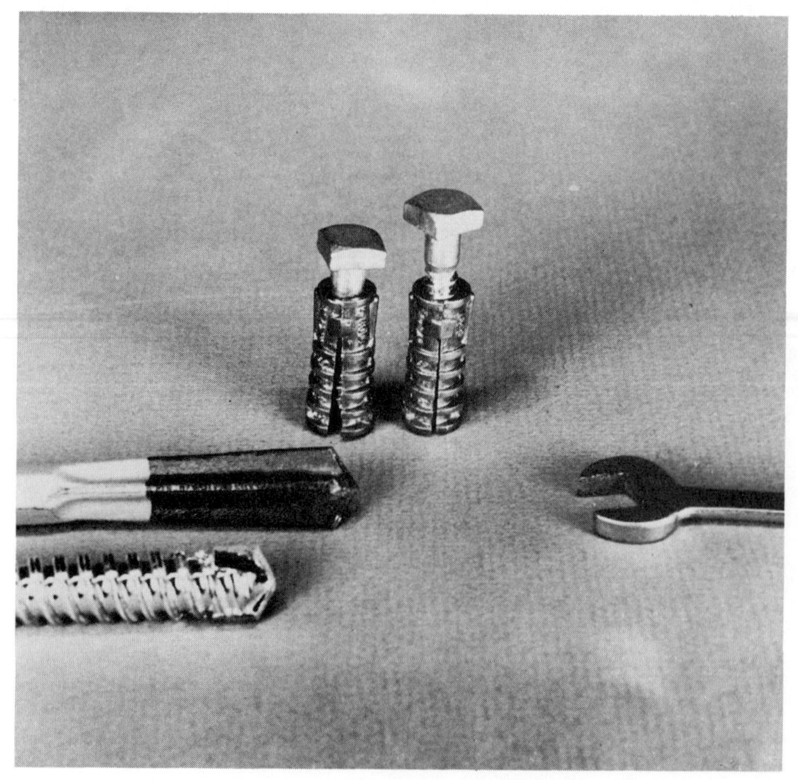

Adhesives are fasteners

Mechanical joints (those using a metal fastener such as a nail or screw) are often made stronger by teaming an adhesive with the fastener; or, the joint may rely on just an adhesive to hold it together. Either way, you should know which adhesive to use for a certain job and how to work with them.

1. When working with wood joints, lightly sand the joints before you apply the adhesive. This will open the pores of the wood, giving the adhesive more holding power. Also, the joints should be square; the more the surfaces touch, the better the grip will be.
2. Clamp glued joints, if possible, but don't clamp the joint too tightly, or most of the adhesive will be pushed out of the joint. As a general rule, tighten the clamp with your fingers only; this supplies about the right amount of pressure.
3. If the joint can't be clamped, use a filler type glue. This will provide a stronger bond.
4. Always follow the manufacturer's directions in mixing glues, and never mix more glue than can be kept during a given drying time. Drying times usually are stated on the adhesive container.

Coat both surfaces of material to be joined with adhesive. This applies to almost all projects. Many adhesives, such as contact cement (used here), dry rapidly, so you have to work fast. And, as a rule, you should work with adhesives in temperatures of 70° or higher.

Picking the right adhesive for the job

Type of adhesive	*Drying time*	*Use*	*Important Data*
Aliphatic resin	Fast	Wood, general	Water resistant; lacquer solvents won't touch it.
Animal/fish	Slow	Wood, paper	Has little moisture or heat resistance. Brittle when dry and aged.
Casein	Medium	Wood, fabrics	Good heat resistance; poor water resistance.
Casein-latex	Medium	Metal, glass, fabric, plastic	Good heat and water resistance; dries clear.
Thermoplastic	Fast	Paper, wood, china, glass	May damage some rubber, plastic, and lacquer finishes.
Thermosetting	Medium	Wood, paper, fabrics	Adhesive won't stain.
Contact	Fast	Joining plastic laminates to wood.	Resistant to heat and water. Work quickly since material dries very fast.
Epoxy	Medium	Wood, china, metal	Expensive; won't shrink; excellent holding ability.

Wood joinery techniques

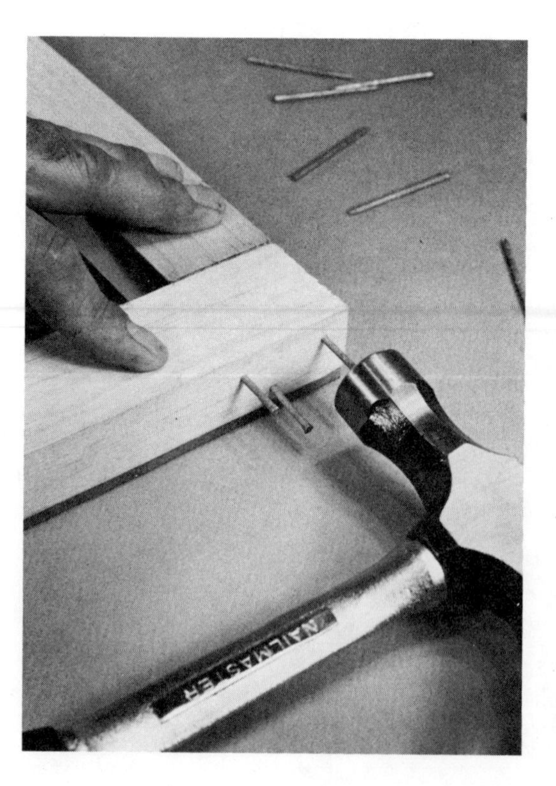

Butt joint. The edges or faces of the stock are joined flush. A butt joint is the basic framing joint; the ends of dimension lumber usually are nailed to the faces of joining dimension lumber. The joint also may be fastened with screws and/or glue. You can give it more strength by inserting a metal angle iron or mending plate along the edge of the stock or across the face of the material. Edge joints (a series of boards are glued together) are also termed butt joints. Support edge joints with cleats (strips of wood or iron for strengthening), if they are in a series.

Although we show a potpourri of wood joints in this chapter, it is recommended that you use the basic ones—at least until your woodworking skills are developed. Fancy joints such as the twin mortise and tenon, double tenon, dovetail, scribed tenon, secret dovetail, and T lap are extremely difficult to make with hand tools. You need power tools for these joints. The joints can be made with hand tools, although by the time you are ready to make them with these tools, you won't need this instruction book.

The basic joints are:

1. The butt joint
2. The miter joint
3. The dado joint
4. The rabbet joint
5. The half lap joint

Even a couple of these basic joints are difficult and time-consuming to make. But there are still others that are more complicated than these.

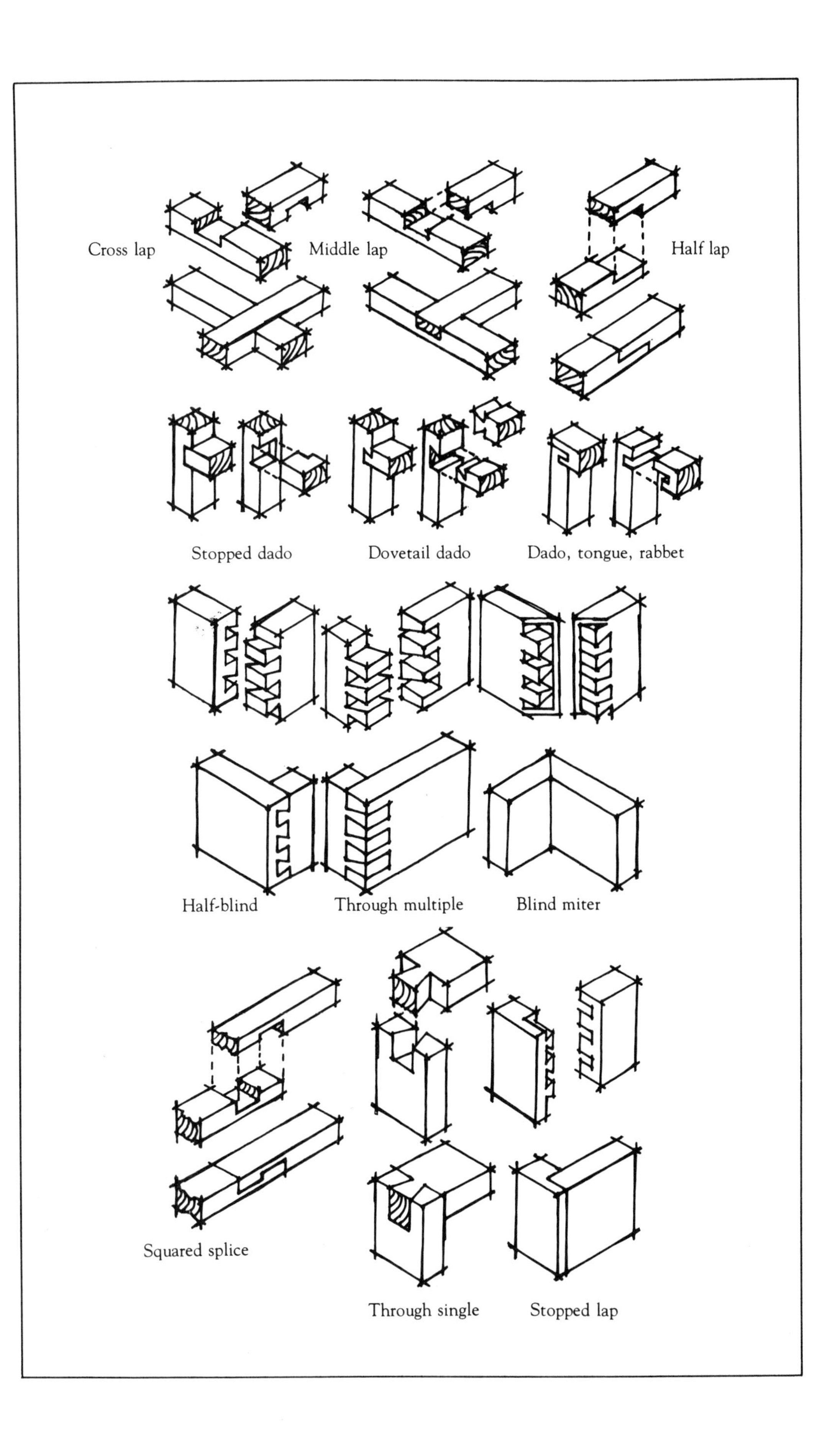
Cross lap
Middle lap
Half lap
Stopped dado
Dovetail dado
Dado, tongue, rabbet
Half-blind
Through multiple
Blind miter
Squared splice
Through single
Stopped lap

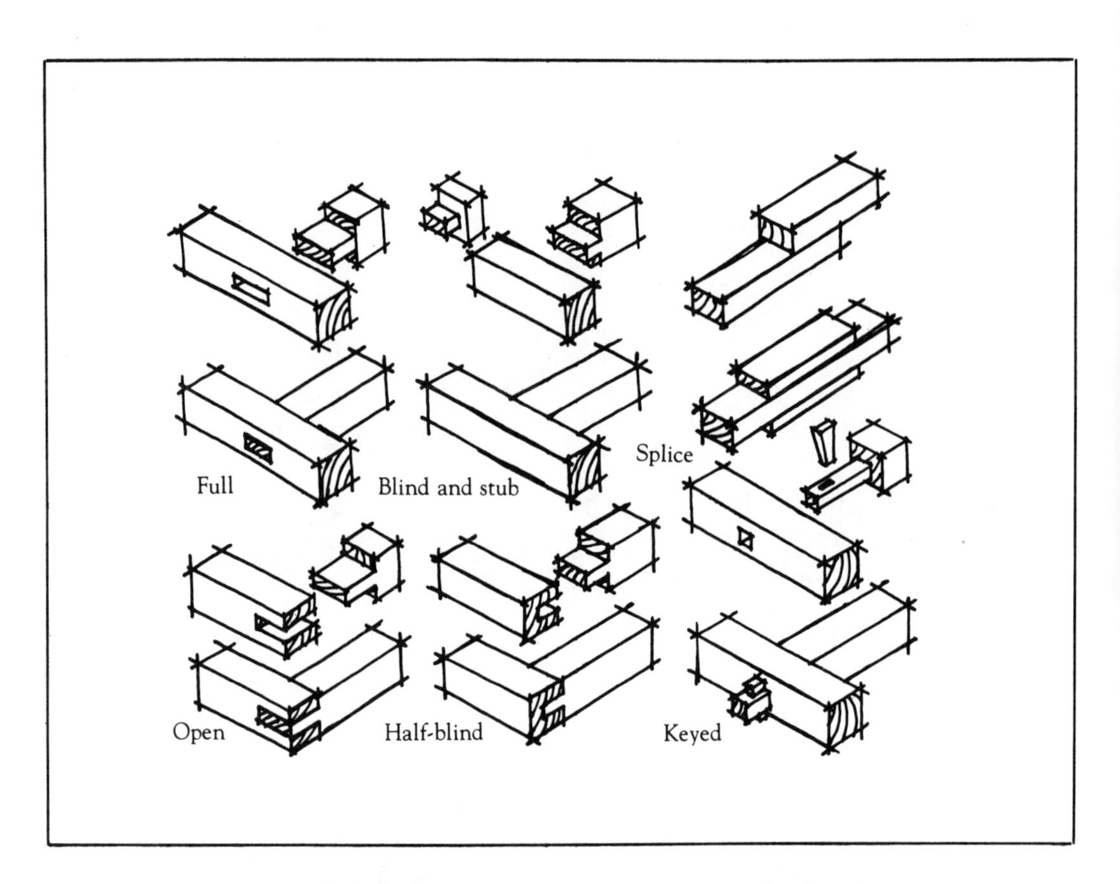

Miter joint. The edges of the stock are usually fastened together at a 45 degree angle. This joint is cut in a miter box with a backsaw. It may be fastened with nails or screws, and is usually countersunk with adhesive for additional support. Or, when using light wood, the joint may be simply glued and clamped until the glue is dry. Also, the joint may be supported with metal angles or mending plates.

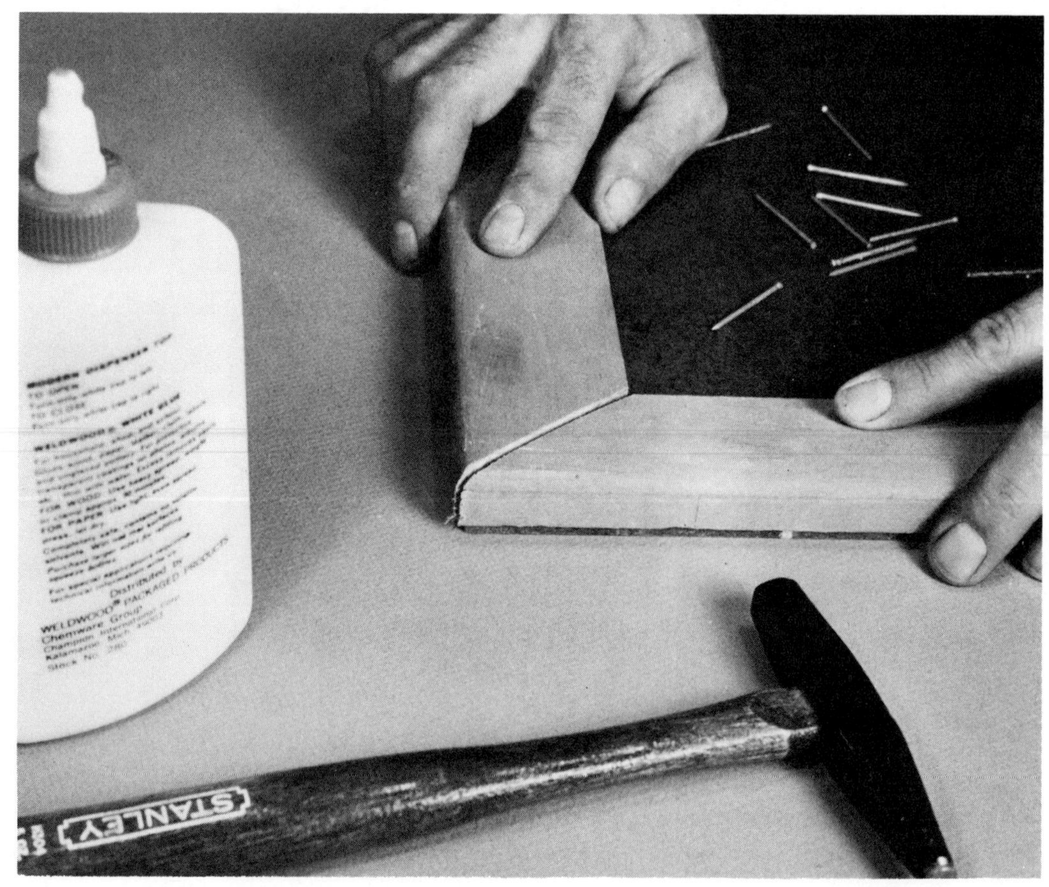

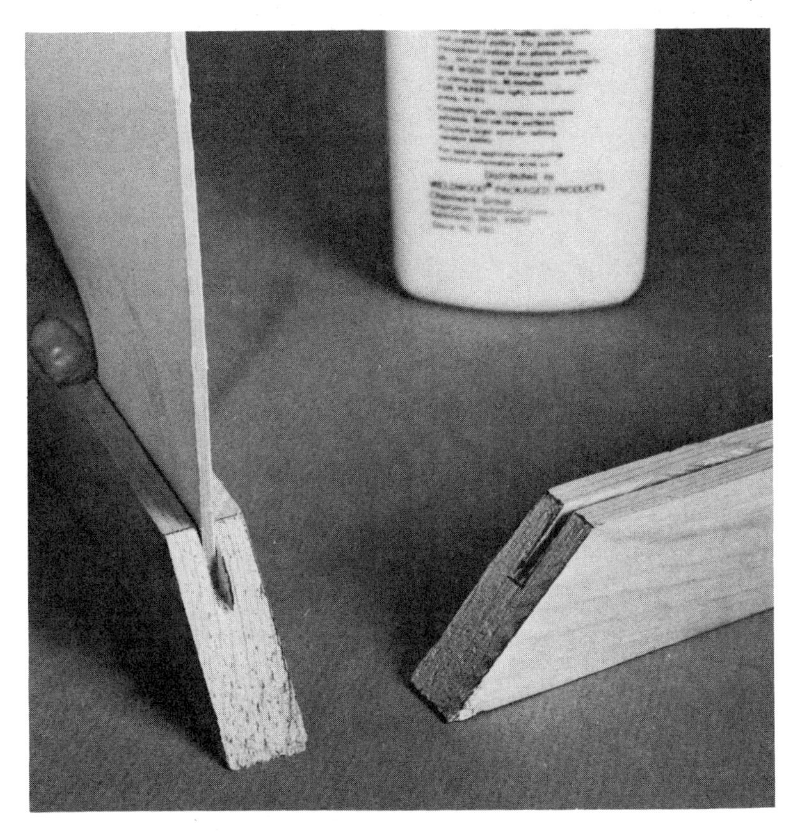

Dado joint. A saw kerf is made down the length of a piece of stock to form a dado joint. Although it is a tough joint to make with a hand saw, it can be done —usually on short pieces of stock. On a power saw, dado joints may be cut by passing the stock over a saw blade that is set at the right height. For wide dado joints, a special dado saw blade is used in a power saw.

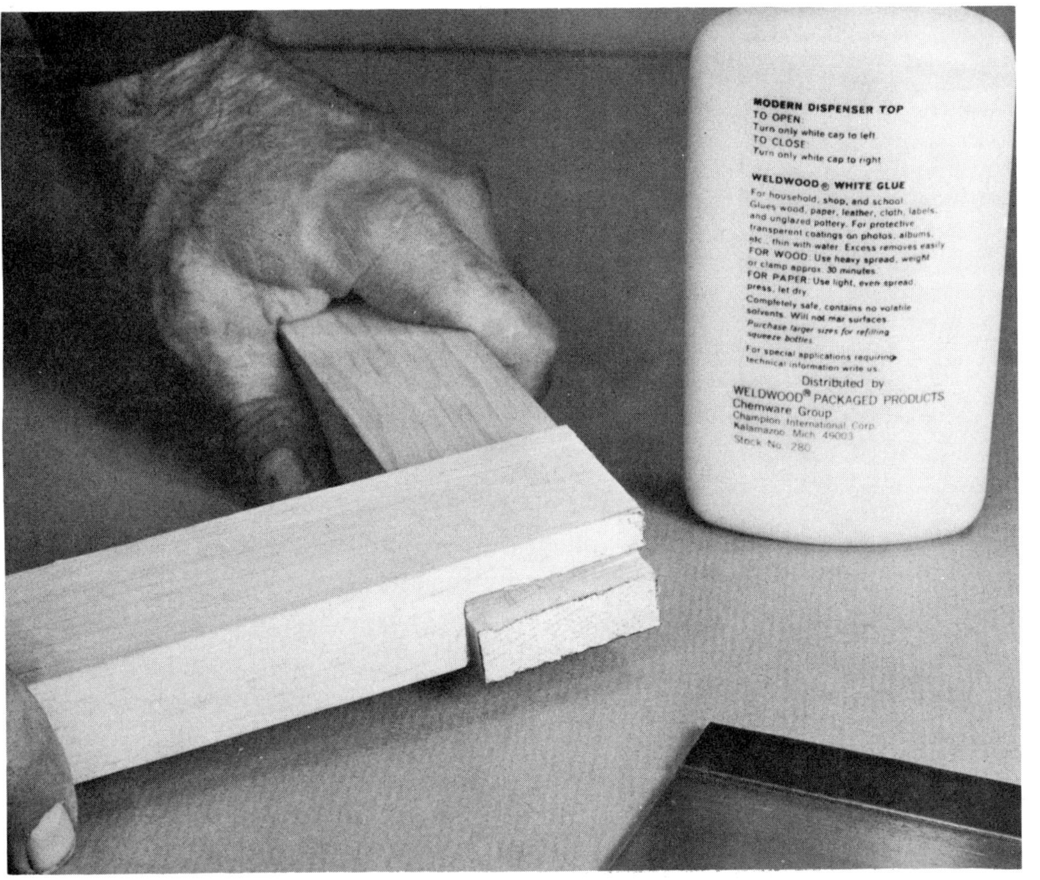

Half lap joint. This is an easy joint to make with a hand tool. Starting at the end of the stock, cut it halfway through its thickness to the depth you want. Then, cut the joining piece of wood the same way. The joint may be nailed or screwed together; also, it may be held with glue or with both metal fasteners and glue. Half lap joints may be used "within" the stock. Here, you make 2 saw cuts the width of the stock; then make a series of saw cuts within these marks. Clean out the excess wood with a flat chisel. Of course, you must make the same series of cuts on the joining piece of material. A half lap is a strong joint to use when sheer weight (weight straight down) is a consideration.

Rabbet joint. Make this joint with 2 saw cuts, one along the face (vertical) of the stock and the other along the edge (horizontal) of it. Or, you can make the joint with a rabbet plane which is especially designed for this job.

Choosing and Using Basic Building Materials

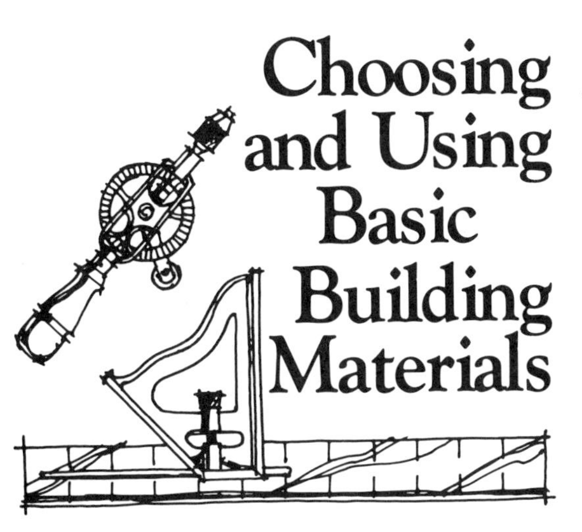

In carpentry and woodworking there are five basic materials: dimension lumber and boards, plywood, hardboards, and moldings. Although not a wood product, gypsum wallboard is another popular building material. Each has its own set of specifications and classifications.

Dimension lumber and boards

There are 2 kinds of dimension lumber and boards: softwood and hardwood. Softwood, such as hemlock, fir, pine, spruce, cypress, and redwood, is used for framing and general building construction. Hardwood, such as mahogany, walnut, oak, and maple, is generally used for cabinets, flooring, fine furniture, and some trim applications.

Dimension lumber refers to lumber that is 2 inches thick in nominal size. For example, 2 by 4's, 2 by 6's, and 2 by 8's are dimension lumber.

Boards are lengths of lumber that are 1 inch thick in nominal size. For example, 1 by 2's, 1 by 4's, and 1 by 6's, are boards. (See the chart in this chapter for standard lumber sizes.)

Lumber grading is varied; grades may be found in numbers, names, and letters. No. 1 grade is construction grade; No. 2 is standard grade; No. 3 is utility grade; and No. 4 is economy grade. Nos. 1 and 2 clear lumber are the best you can buy. B and better grade lumber is permitted to have some slight imperfections. C select grade has limited imperfections. D select grade has many imperfections, but they may be hidden with several coats of paint. Common lumber may have weaknesses, while select lumber usually is sound and strong.

Lumber is sold by the board foot. A board foot is a piece of material that is 1 inch thick, 12 inches wide, and 12 inches long. Don't worry about computing the board feet, since the lumber dealer will compute it for you. Just tell the salesman how long and how thick you want the board to be. Lumber and boards are often sold by the piece; these items are usually stamped with the price.

Plywood

Plywood sheets are available in 2 types: hardwood-faced plywood and softwood-faced plywood. Softwood faces are commonly of fir; hardwood faces include such top veneers as walnut, birch, maple, and cherry. Plywood is subjected to industry grading standards which assure you that material purchased is uniform. There are standards that permit plugging knotholes and mending split voids. This doesn't mean that the material is inferior quality; instead, it assures quality.

Hardwood-faced plywood grading systems differ from the softwood grading systems. Custom grade No. 1 is free of knots, patches, or plugs. Good grade No. 2 is sound, with evenly matched veneer. Sound grade No. 3 will have some defects; the veneer may not be evenly matched, and there may be some mineral streaks in it. Utility grade No. 4 may have discolorations in the veneer. Reject grade No. 5 generally will have knots and splits.

You can buy grade combinations of hardwood-faced plywood. For example, if you want both sides to be good, buy 1-1; if you want only one good side and the back will not show, buy 1-5.

Also, the glue bonds of hardwood-faced plywood are classified as waterproof, water resistant, and dry. For instance, if water is a problem, you would buy waterproof bond. If the panels are going on a living room wall, your selection would be dry.

Generally, softwood plywood is made of fir, larch, pine, loblolly, and longleaf pine. The panels are graded as N, A, B, C, C-plugged, and D.

N-grade is a select panel. It may be all heartwood or all sapwood, and it is usually free of any open defects. This grading does permit some repairs (the mill plugs defects with solid wood material).

A-grade may have some repairs. It is smooth and sanded, and may be stain finished.

B-grade permits repair plugs and tight knots.

C-grade permits knotholes, limited splits, and minimum veneer on some exterior types.

C-plugged has splits, knotholes, and borer holes; the size of the holes is limited.

D-grade has knotholes. Grading permits limited splits in the material.

Like the hardwood-faced plywood, the faces of softwood plywood can have two grades. You can buy an A face with a C back; or you can buy a C face with a D back.

An example: If you want to buy a piece of plywood with top grade veneer on both front and back sides, you would ask the lumber dealer for A-A grade. Since you probably are going to use this material inside, you would ask for interior type.

Plywood thicknesses commonly are ¼, ⅜, ½, ⅝, ¾, and 1 inch. Veneer grades are A face, A back, and D-grade inner plys. Sheets measure 4 by 8 and 4 by 10 feet. Plywood may be purchased in decorative panels that have been brushed, grooved, striated, embossed, or rough sawn on one side. Also, you can buy special panels for building boats, such as 2-4-1 panels for subfloors and underlayment; overlayed plywood is used for exterior cabinet projects, windscreens, siding, and soffits.

Hardboard

Hardboard panels are really wood; the material is made from wood chips that are turned into sheets under heat and pressure.

There are 2 types of hardboard: standard and tempered. However, there are variations of these types. Generally, standard hardboard is recommended for interior applications where moisture is not a problem. Tempered hardboard has been specially treated to withstand moisture; it may be used outside and in an area such as a bathroom where high humidity may be a major problem.

Also, you can buy black tempered hardboard for a flat black finish; this panel hardboard may be used for walls, soffits, and ceilings. It is not as "hard" or "dense" as regular tempered hardboard. Underlayment hardboard for floors provides a smooth surface for tile and carpeting. If both sides of the panel will be seen, you can buy hardboard that is smooth on both sides. And, you can buy literally hundreds of hardboard panels in a variety of designs; embossed, perforated, concrete form, and sidings.

Sizes of hardboard panels range from 4 by 4 to 4 by 16 feet, with the standards set at 4 by 8 and 4 by 10 feet; thicknesses are 1/8, 3/16, 1/4, and 5/16 inch.

Hardboard manufacturers also have special trim moldings available for paneling projects. These include snap-on metal moldings for vertical joints, inside and outside corners, and cap moldings. Or, you can use regular wood moldings such as batten strips, covers, and others.

Moldings

Molding patterns are so varied that it would take a special book to list them all. However, the basic ones are: crowns, coves, quarter-rounds, base shoes, half-rounds, drip caps, corner guards, rounds, squares, pictures, lattices, mullion casings, ply caps, and shingles.

Moldings are sold by the linear foot, and the pattern of the molding will determine the price. For example, cove molding is less expensive than brick molding because of its mass and configuration.

Gypsum wallboard

Gypsum wallboard is often used in remodeling and in new building projects. The wallboard has a gypsum core with a paper covering which is fire resistant, and it is also a replacement for a lath and plaster ceiling and/or wall treatments.

Standard gypsum wallboard is made in ⅜, ½, and ⅝ inch thicknesses. Panels range in size from 4 feet wide to 16 feet long; standard sizes are 4 by 8 and 4 by 10 feet.

Panels may be purchased with a slight taper on the long edges; the taper permits the joint tape to fit flush over the joints of the material. Other edge treatments include round, beveled, eased, square, and tongue and grooved. And, you can buy specially treated papers; some are decorated, and others are coated with moisture-resistant sealers.

If you use ⅜- or ½-inch gypsum wallboard, you will need 1⅝-inch gypsum wallboard nails; if the thickness is ⅝ inch, use 1⅞-inch nails.

Used lumber is a bargain

Don't overlook used lumber for a building project, especially if the lumber is used for framing and will not be seen.

Tips on working with plywood

Plywood is easy to saw and plane, but there are several rules you must follow for a smooth, easy job.

1. Draw multiple pieces to be cut out of 1 sheet of plywood on a sheet of paper, thus laying out the job. This will prevent an error in cutting, and it will also save money later. Transfer the dimensions to the plywood following the paper pattern.
2. Use a crosscut saw to cut plywood. Saw with the best face of the panel up, and be sure to properly support the panel so that the saw doesn't bind.
3. Cut with the best side of the panel down, if you use a power saw. Use a combination blade.
4. Plane plywood from the ends toward the middle; this will prevent splitting corners and edges.
5. Sand with the direction of the grain, using a medium-grit abrasive. Go easy to keep from cutting into the veneer with the abrasive. Always use abrasive on a flat sanding block.
6. Avoid nailing or screwing into the edge of veneer core plywood; fasteners can split the veneers.
7. Seal edges of plywood with a clear penetrating sealer. If you sand the face of the material and the back, give both surfaces a coat of sealer.
8. Read the painting and stain schedules on finish containers for the correct application. Plywood will take almost any kind of finish. For the "wild grain" in fir plywood, use a clear penetrating sealer before the final finish is applied to the surface.

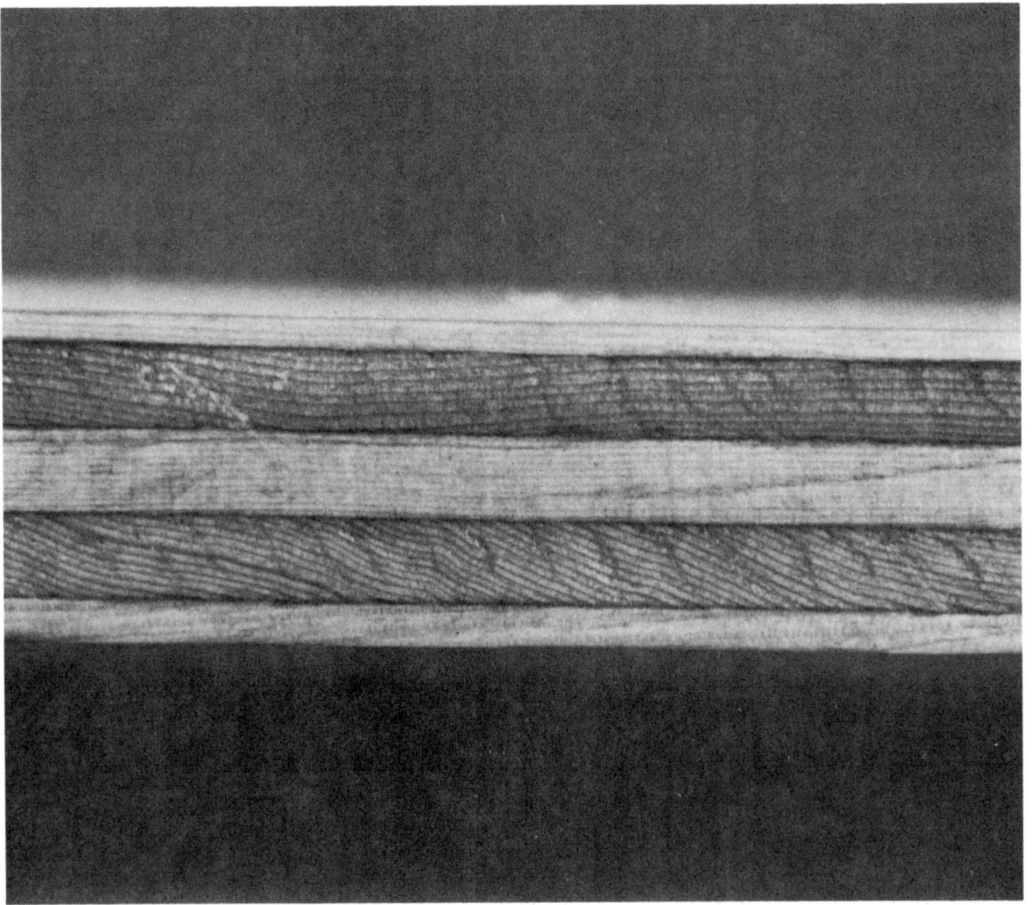

Veneer-core plywood has 3 inner plies with top and bottom veneer fastened to them. The veneers go at right angles to each other in an alternating pattern. As a rule, the number of plies determines the strength of the plywood. Exterior type plywood has a waterproof bond; interior type plywood has a water-resistant bond.

Lumber-core plywood, as the name implies, is sheets of veneer laminated to a core (middle section) of solid wood. Lumber-core plywood is especially strong since the core is edge glued and fitted for stress. This plywood is very suitable for cabinetmaking.

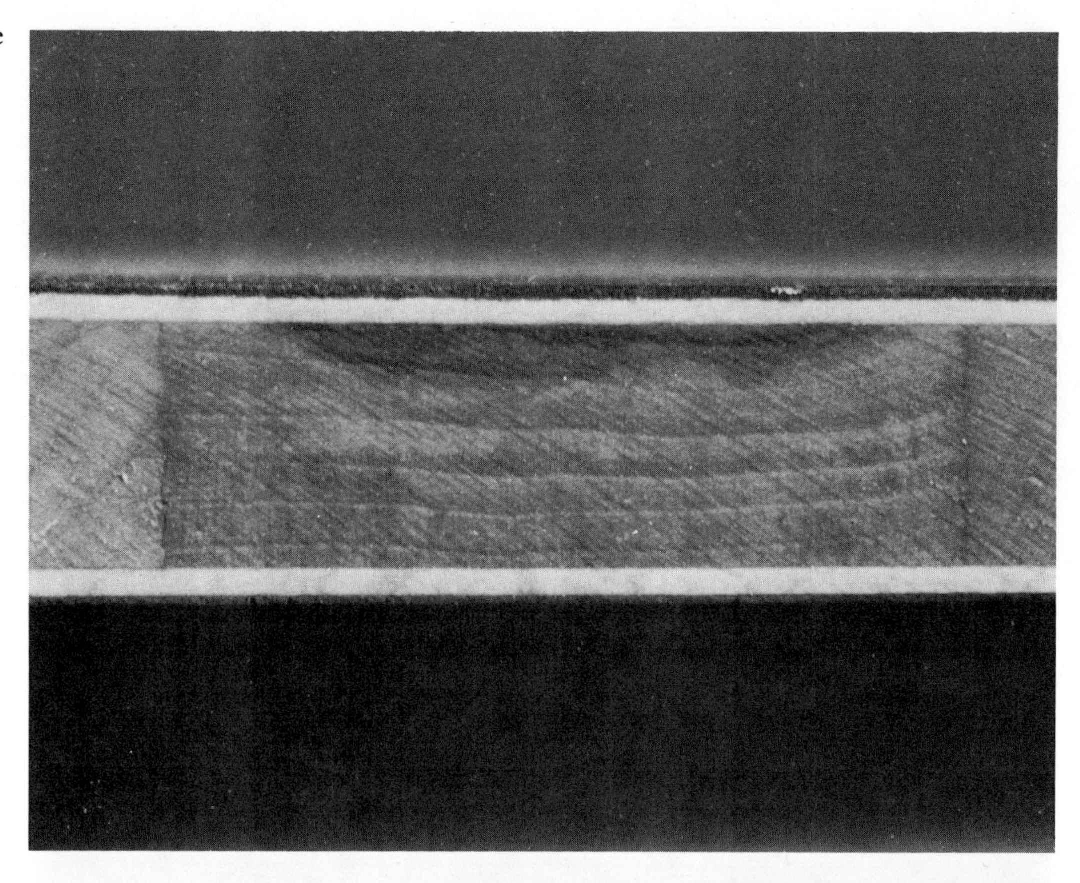

Particle board core plywood has core fabricated from tiny particles of real wood that have been resin-coated. Particle board also is available without the veneers bonded to the smooth face surfaces; it may be used for some cabinets, built-ins, and as a floor underlayment.

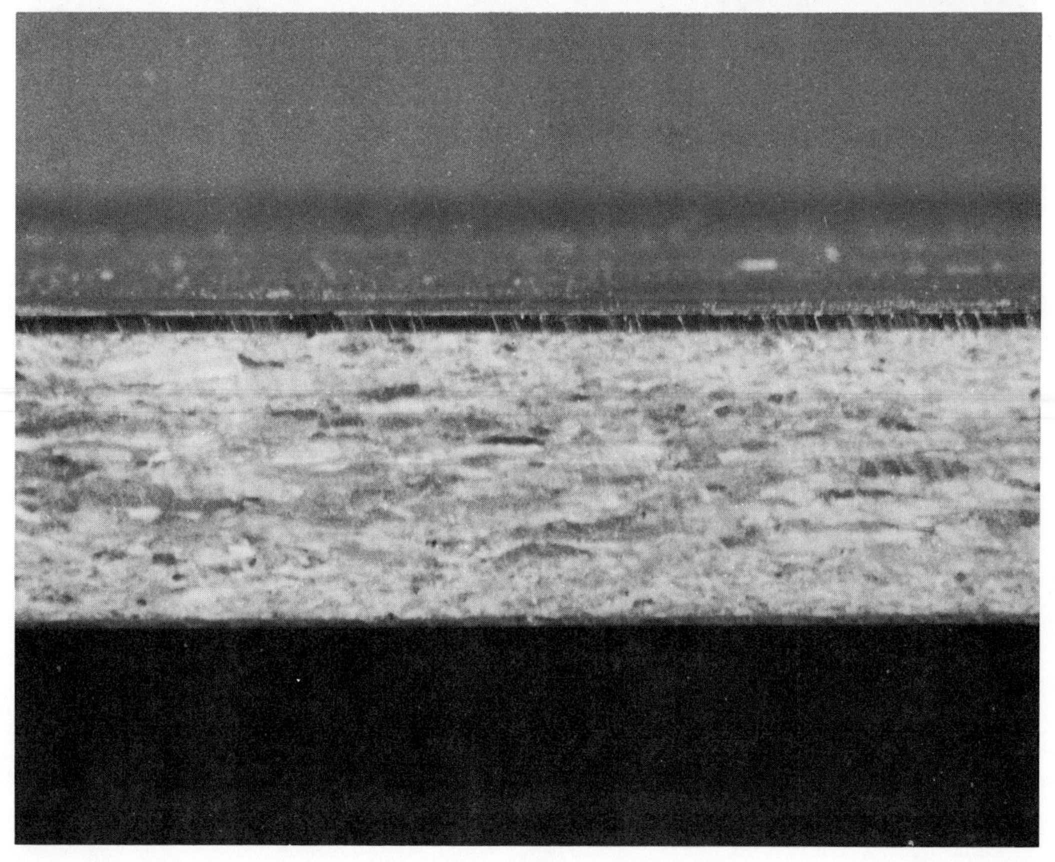

Tips on working with hardboard

Hardboard is tough and your cutting tools have to be sharp. However, you don't need special tools when working with the material.

1. Use a combination or crosscut blade for sawing hardboard. For multiple cuts, use a carbide-tipped blade on a power saw. Hardboard may also be cut with a coping saw, a keyhole saw, or a hacksaw.
2. Drill holes with twist drills and work from the "finished" side of the sheet.
3. Hardboard is "grainless." You can nail or screw hardboard to another surface, but you can't fasten another material to it.
4. Hardboard must have solid support. If the job calls for ⅛-inch-thick material, make sure it is supported continuously. If the material is thicker than ⅛ inch, you should support it on furring strips, studs, joists, rafters, and other types of framing—just so the spacing of the framing is not more than 16 inches on center.
5. Don't push the hardboard panels tightly together. The material absorbs some moisture, which will cause the panels to warp. Leave a little space for expansion and contraction. The joints may be covered with a molding or a batten strip.
6. When hardboard (and plywood, too) is installed over surfaces that may become damp at times, such as foundation walls, cover the walls first with a moisture vapor barrier. Polyethylene film makes a good barrier.
7. You can use contact cement to fasten hardboard to large wall, ceiling, and floor surfaces. A special paneling adhesive is made for this application.
8. Hardboard may be finished with any type finish that is recommended for regular wood surfaces. As a rule, the surface has to have a prime coat of finish and then one or two additional coats of finish.

Nominal sizes of lumber and boards are not the actual sizes. This is a piece of 5/4-inch lumber which actually measures about 1 inch thick, not 1¼ inches thick.

Nominal sizes of boards are 1 by 2, 1 by 3, 1 by 4, etc. Actual sizes are ¾ by 1½, ¾ by 2½, and ¾ by 3½. This is a piece of 1 by 3 stock which measures ¾ inch thick by 2½ inches wide.

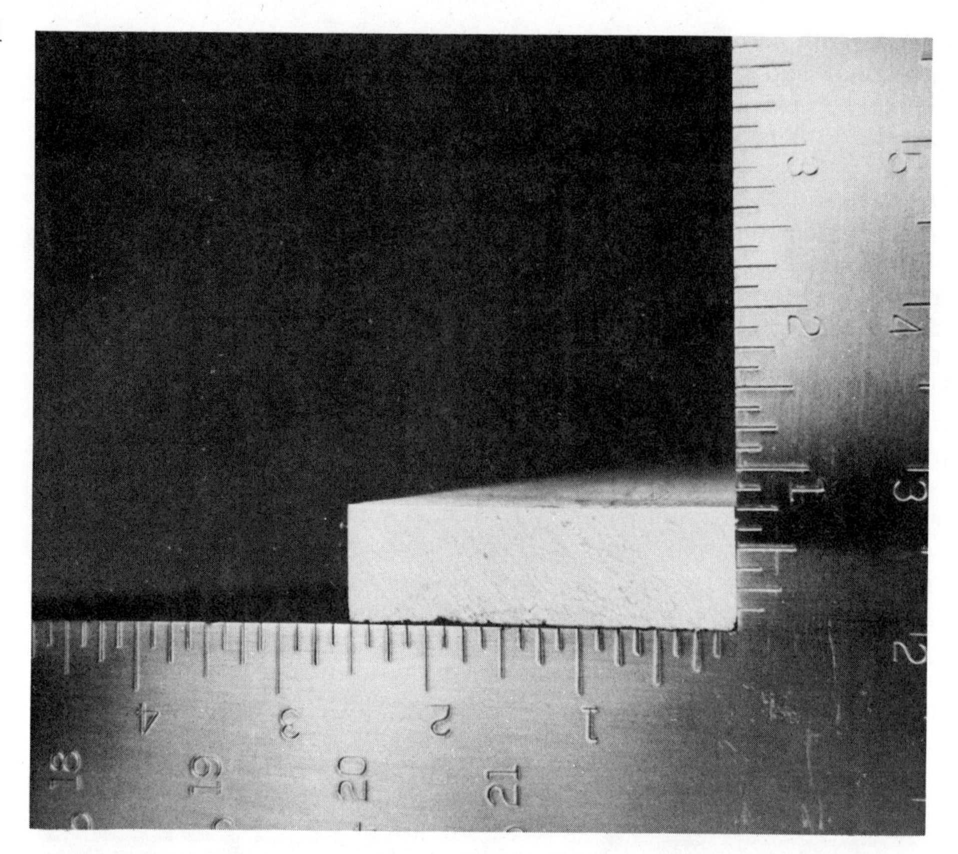

Dimension lumber has a nominal size standard, too. This is a piece of 2 by 6 which actually measures 1½ inches thick by 5½ inches wide.

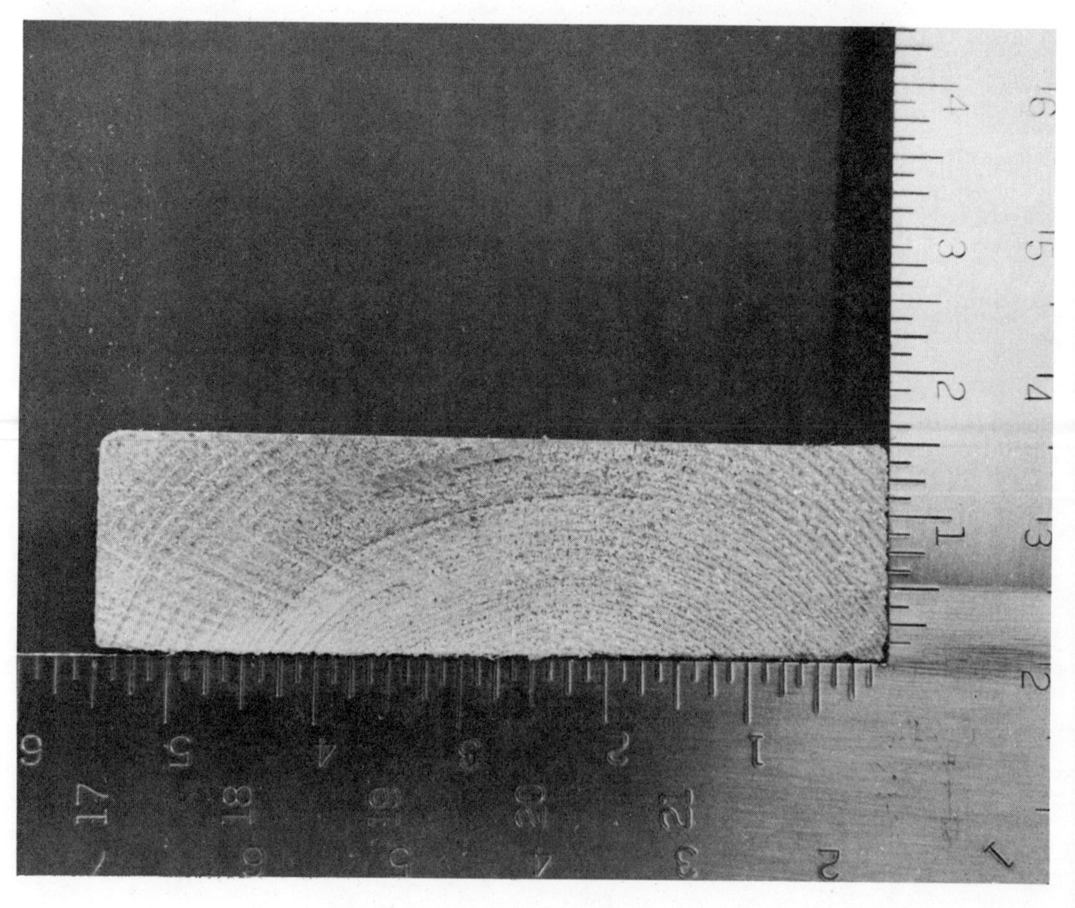

Basic lumber sizes

Lumber type	*Nominal size (inches)*	*Actual size (inches) (Material surfaced dry on four sides)*
Boards	1x2	¾x1½
	1x3	¾x2½
	1x4	¾x3½
	1x5	¾x4½
	1x6	¾x5½
	1x7	¾x6½
	1x8	¾x7½
Dimension	2x4	1½x3½
	2x6	1½x5½
	2x8	1½x7¼
	2x10	1¼x9¼
Timbers	5" thicker	½ off nominal
Shiplap	1x4	¾x3⅛
	1x6	¾x5⅛
	1x8	¾x6⅞

The raw edges of plywood may be covered with matching veneer tape (as shown) or any number of moldings or splines. If you use veneer tape, the edges must be sanded square and smooth. Seal the surface with a clear penetrating stain; then apply contact cement to the back of the tape and the edge of the plywood. Wait until the cement is dry to the touch, then carefully stick on the veneer tape. Once the tape meets the edge surface, it forms a very strong bond which is almost impossible to pull loose and reset.

Typical molding configurations are (from left) the quarter round base shoe, crown bed, casing base, drip cap, and crown bed. Some moldings may be purchased prefinished, especially those that are applied to paneling installations. Standard wood moldings are finely sanded; then they are ready for paint or stain finishes.

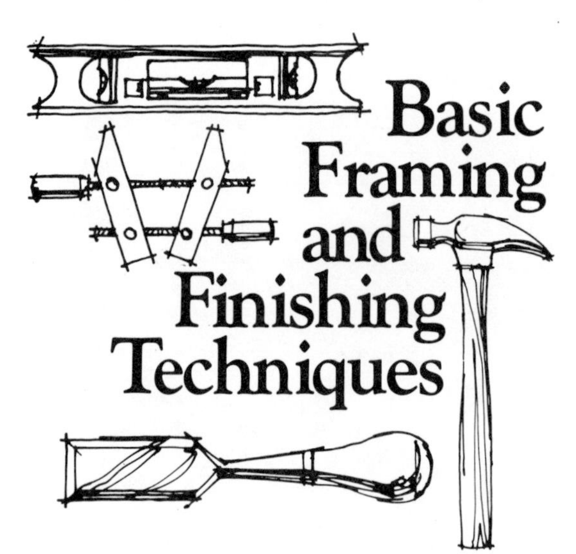

Basic Framing and Finishing Techniques

Whether you're building anew or reshuffling the old, there are certain basic framing and finishing techniques you have to follow to do the job properly.

The illustrations in this chapter are intended to provide you with patterns to copy when you decide to add on a new room, put a new door or window in your existing home, or add a new partition wall in the laundry or family room. Most framing or finishing situations are covered in this chapter; however, if you have a special problem, your building material retailer can probably help you.

You will need a building permit before you can begin a building project. And, most projects have to comply with local building codes; make sure you meet these requirements before you start pounding away. Check with local building code officials or your material supplier to prevent future problems and to save money at the outset. Also, have a plan for the project; for simple construction, such as a partition wall, you can rough out a drawing yourself. For a project such as a room addition, you should consult an architect or engineer.

Most framing members are set on 16-inch centers. An exception would be a garage framing, which sometimes is put on a 24-inch center. Almost always, the material used for framing members is dimension lumber—2 by 4's, 2 by 6's, and 2 by 8's. Studs usually are 2 by 4's, as shown.

Most framing members fit together in butt or edge joints. The ends of the members must be square and fit tightly together. The joints are toenailed together; the nails are slanted or angled through one piece of wood into the other piece of wood.

Houses and additions sit on foundations. The foundation may be concrete block or reinforced concrete. When using either foundation, the bottom has to be below the frost line to prevent it from heaving and breaking. Check your local building department for frost line depths. It usually is best to have a professional place the foundation for you, since the job will take special digging equipment and techniques. For example, the concrete forms here are plywood locked together with metal braces and special screw hooks.

Reinforced concrete foundation looks like this when the forms have been stripped. The sill bolts hold the sill to the foundation; a junior I beam, formed from steel, supports the middle weight of the addition or house. It sits on a concrete pad for support.

Notches are cut out of sill headers so that sill bolts may be tightened before the final finishing materials are added to the structure. A partition wall stud above the floor underlayment has a 2 by 4 stud which is turned edgewise. This is to enable the adjoining wall framing to butt against the stud, forming a strong partition corner.

Sills are shimmed level with cedar shingles. Then the sill bolts are tightened, and the sills are again checked for levelness. Sill bolts are placed from 24 to 36 inches apart, and the sill nuts are supported by washers.

Styrofoam insulation is sometimes used between an interior foundation wall and a poured concrete floor. The thickness of the insulation depends on local building codes, but it's usually ¾ inch thick. Concrete will cover the insulation when the floor is laid.

Partition walls that are load-bearing are usually set on concrete footings. A load-bearing partition supports weight from the roof structure. The partition sill or plate is bolted into position, and the sill bolts are set when the footings are placed.

Insulation board strips or expansion joint materials are sometimes placed between the partition wall sills and the floor, if the floor will be of poured concrete. This permits expansion and contraction between the two materials. If the wall ever has to be moved, the strips will make it easier to remove the wooden sill.

Heating ducts are installed below a poured concrete floor (as shown). The long lengths of pipe are made of fiber material; the joints are galvanized steel pipe. The pipes, sill insulation, and foundation insulation will be covered when the concrete floor is placed.

The corners of outside walls are supported with plywood sheathing, while the spaces between the corners are filled with insulation board sheathing. Insulation board has a greater resistance to heat and cold than plywood; depending on the climate, regular insulation may not have to be used in conjunction with insulation sheathing.

Window placement in outside walls may call for more framing support. To do this, double 1 stud placement with 2 studs and a 2 by 4 spacer. The load-bearing wall is now stronger, giving the proper roof support.

The support for a partition wall, until the wall is properly fastened, is a 2 by 4 spiked to an outside wall, joist, or rafter. The support holds the wall square for nailing and insures a tight fit against the adjoining framing.

A corner framing joint looks like this. One 2 by 4 member is edge-nailed against a flat 2 by 4 member. The flat 2 by 4 member of the adjoining wall is then nailed to the other 2 by 4 member. The sheathing is then overlapped.

Bearing wall headers or plates are double 2 by 4's that are spiked together. The corner support is also doubled to handle the weight of the roof structure. The headers are spiked or toenailed together where they meet at the corners of the wall sections.

Non-load-bearing partition walls are fastened to the subfloor with nails. In existing buildings with a concrete floor, the bottom sill is fastened to the floor by drilling holes for lead expansion anchors. Insert the anchors into the holes, and fasten the sill with lag bolts that are driven into the anchors.

Closet framing consists of 2 by 4's on 16-inch or narrower centers. The top of the door has a sleeper framing; the wall covering simply goes over these short studs, which forms a nailing surface. The jambs (vertical supports) are double 2 by 4's; the header is a single 2 by 4.

Steel junior I beams and posts, in a post and beam framing configuration, are often used to support wide spans—here across the rear section of a 2-car garage. The posts usually require concrete footings for support.

The plumbing runs go between the floor joists in 2-story construction and in some 1-story plans, when basement or crawl space is provided. Height of the ceiling should be considered, since some piping has to go below the joists. A ceiling treatment (usually a soffit) hides the pipes in living areas.

Joists are lapped (as shown) along long floor spans. The weight is supported by a load-bearing wall, framed with double 2 by 4 headers or plates. The joists are spiked together to prevent any major movement.

Plywood spacers can be used to distribute the weight load on framing members or to make them wider to accept finishing materials. Nail through the framing and the spacer to hold the spacer in position.

Exterior doors have double headers and double jambs. One jamb is cut short so that it will support the headers. A stud is sometimes used next to the jambs to help support the weight from above and to provide a nailing surface for the wall covering—gypsum board or wall paneling.

Window headers on outside bearing walls are usually 2 by 8's or 2 by 10's; they are set on edge and spiked together. The jambs are double 2 by 4's, and one 2 by 4 supports the header on each side of the window opening. Use prehung windows in openings like these. You can get the framing measurements for prehung windows from the lumber dealer where you do business. Most prehung windows are in standard sizes.

Window sill framing may be a single or a double 2 by 4, supported by a vertical stud nailed to the sill of the wall. The studs between the jambs and at the sides of the jambs offer support and provide a nailing surface for the wall covering.

The finished rough opening for a prehung window unit includes: double headers, double jambs, sill support members, and "sleeper" studs which are located under the window sill for the necessary support. Asphalt building paper is tacked around the window as a moisture and insulating barrier. The prehung window is nailed to the framing in the rough opening.

Bow and bay windows also may be purchased as a single, prefabricated unit. The header and jamb framing on these units is very similar to the regular prehung windows. The blocks under the window offer support and a nailing surface for the wall covering.

The roof rafters, which are cut at an angle, are spiked to the ridge support. The angle depends on the pitch of the roof you want. A bevel gauge or rafter square will help you to mark the correct angle for sawing the rafter pieces. The bottom or horizontal pieces of lumber are ties that are nailed to the bottom of the rafters. These members also offer a nailing surface for the ceiling wall treatment, which may be fiberboard tile.

Using prefabricated roof trusses is the easiest way to frame a roof structure. You can buy them ready-made at many building material outlets. The trusses will save a considerable amount of cutting time, since they are simply tipped into place and nailed. Heavy steel plates are used to hold the truss members rigidly together.

Ends of roof trusses are fastened to the plates or headers of the bearing wall sections. Once in place, the roof trusses are covered with sheathing or roof decking, depending on the design of the house.

Metal hangers are sometimes used to support trusses and floor joists—especially when nailing in tight quarters is a problem. The hangers are face-nailed to the support members. The joining material is then nailed to the hangers to prevent them from moving.

Vertical plumbing is installed between the studs. Or, the studs are drilled at the center to allow the plumbing to be threaded through the holes. An electrical conduit is also threaded through holes drilled through studs, joists, and other framing members.

Insulation board sheathing and plywood sheathing are nailed to the studs, sills, and headers on outside walls. Nailing positions for this material are usually furnished by the manufacturer; some sheathing is even marked for nailing.

Soffits for roof overhangs may be purchased prefabricated or you can frame your own with plywood or hardboard. The ends of rafters may provide nailing supports for the studs to which the soffits are fastened. Or, you can build a U- shaped box, using a ledger strip (horizontal board) along the sheathing for nailing the back side of the soffit board. The facing or trim board is then nailed to the soffit.

Cut gypsum wallboard with a sharp razor utility knife. Score the paper covering with the knife. Then lay the material over a piece of 1 by 4 or 2 by 4 so that the edge of the wood follows the cutting line. Snap the wallboard over the edge of the wood. The wallboard will break evenly along the scored mark. Then cut the paper on back of the board to complete the job.

Split the stud with the wallboard for nailing. One-half of the stud is covered with one piece of wallboard; the adjoining half covers the rest of the stud. Use regular gypsum board nails, spacing them about 3 to 5 inches apart.

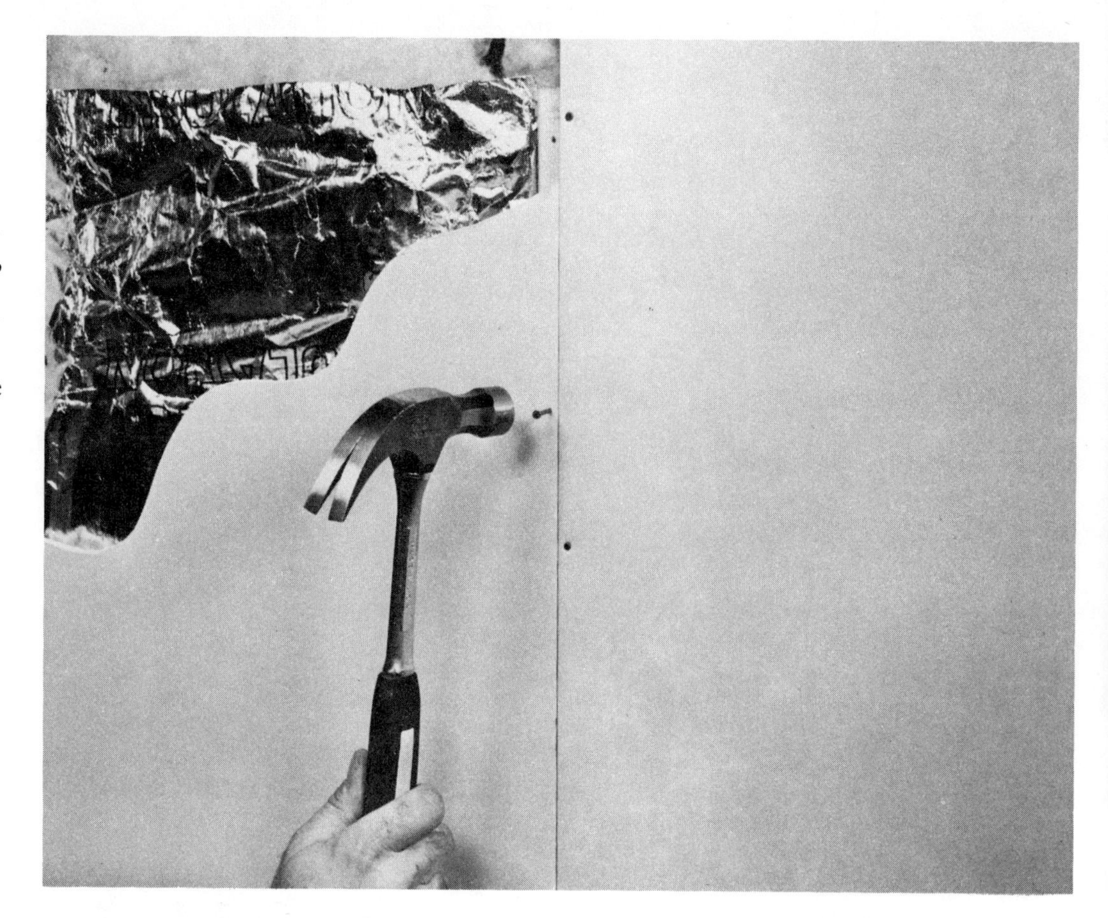

Stagger the gypsum board nails on each side of the joint. Drive the nailheads flush with the wallboard, then hit the nails once more. This will leave a slight "dimple" in the surface of the wallboard. The dimple will be filled with joint cement, hiding the nailheads so they won't show through the cement and tape.

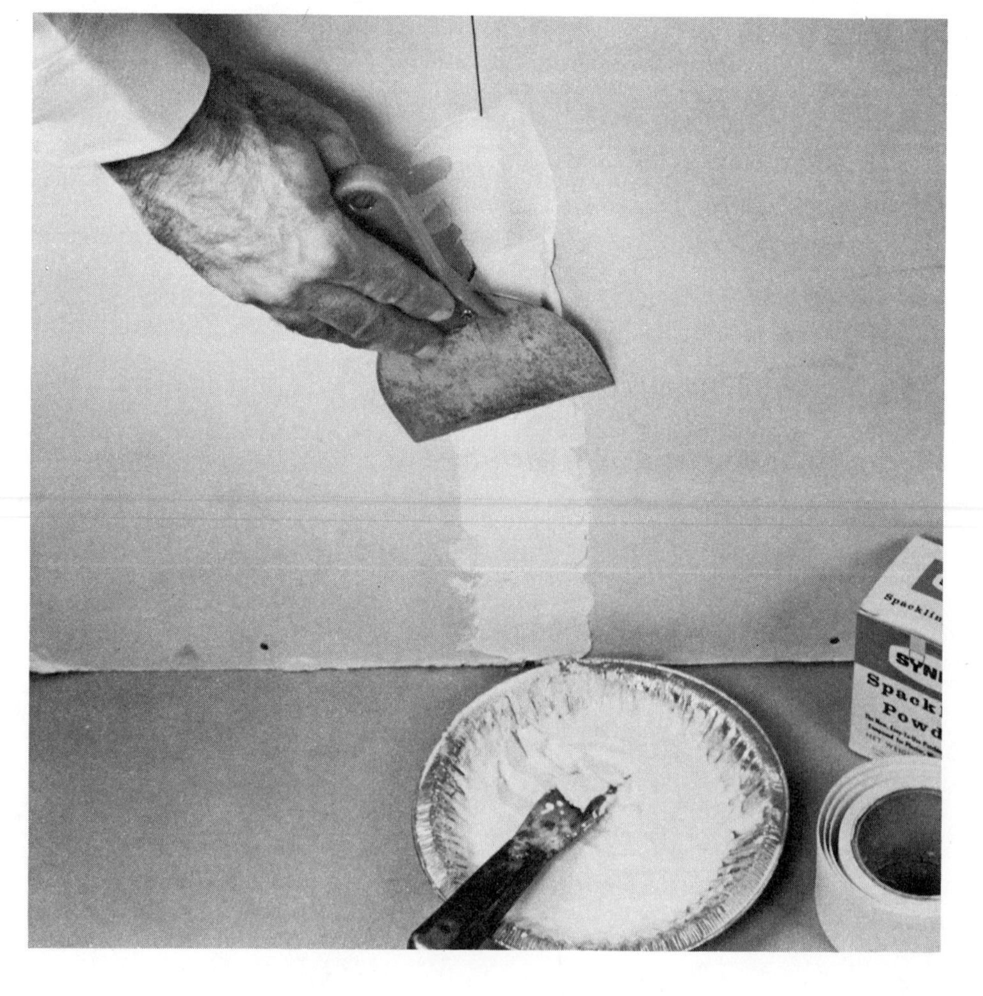

Spackling compound or gypsum wallboard cement is used to cement the tape into the joint. The cement is mixed to the consistency of thick whipped cream. Apply the cement to the joint with a wide wall scraper. Be liberal with the cement to fill up the joint.

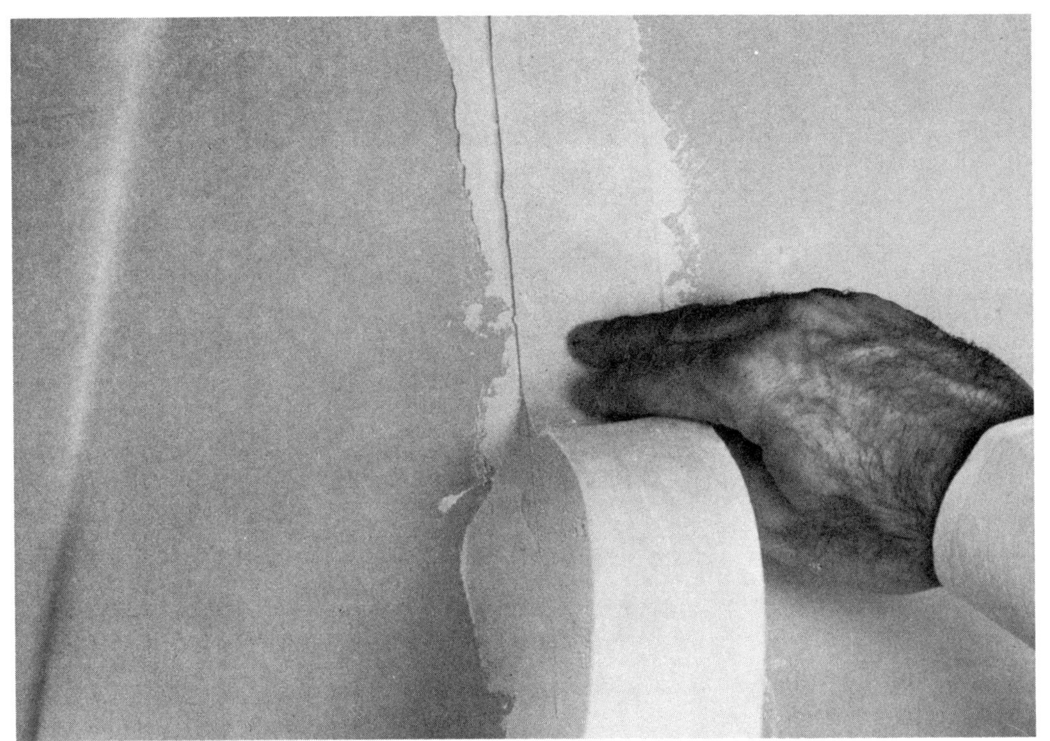

The joint tape is imbedded into the cement base. You may have to reposition the tape several times to make it smooth on the joint. For inside and outside corners, a metal corner molding is used. This is nailed to the gypsum board. Then the edges of the molding are cemented and covered with tape similar to that used on regular wallboard joints.

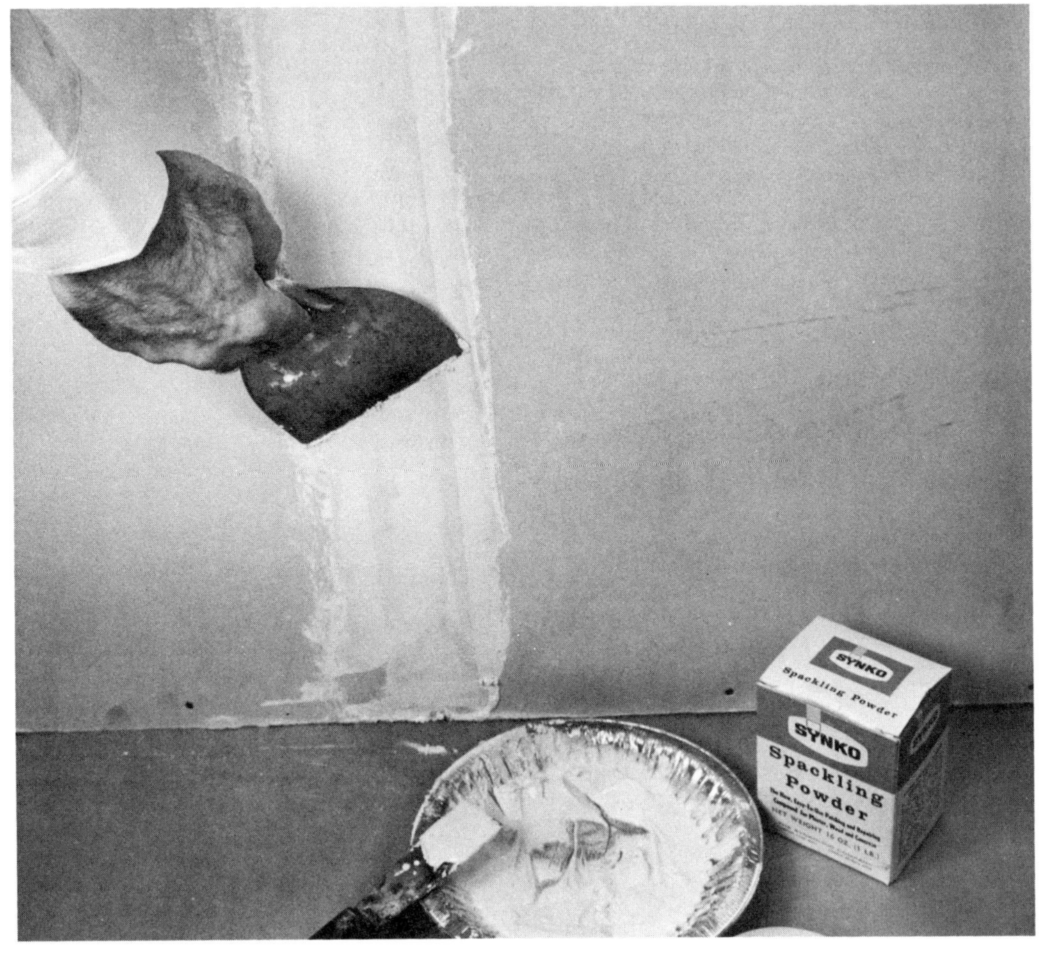

Press the joint tape into the compound with the wall scraper. Then apply another thin coating to the top surface of the tape. Run the scraper over the compound to make it as smooth as possible; do not leave any of the tape exposed. If it is buckled or bunched over the joint, remove it and try again.

Sand the joint smooth when the joint compound is dry. If you have the time, let the job set overnight. Use medium-grit sandpaper stretched over a flat sanding block. Don't dig or gouge the wall; the cement is thin and scars easily. The wall is now ready to be finished with paint, wallpaper, paneling, or wall covering.

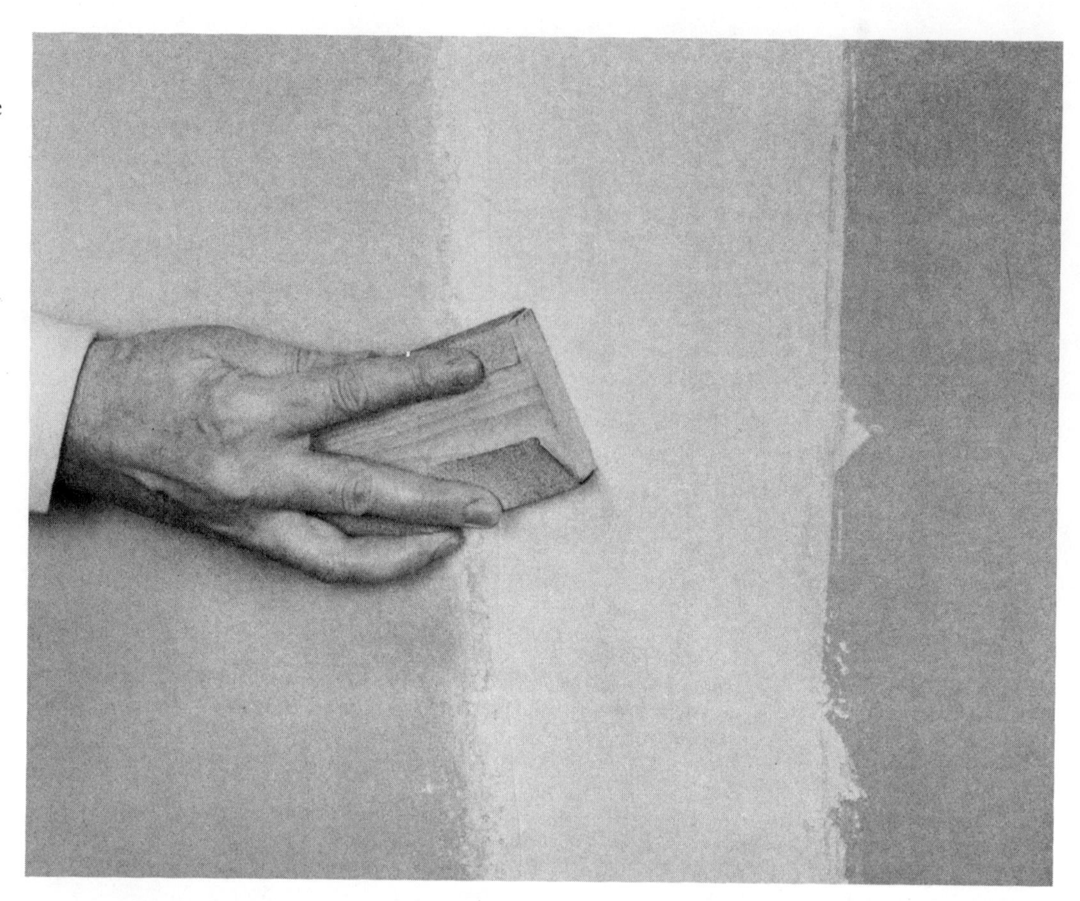

Furring strips for ceiling tile and paneling

Ceiling tile and wall paneling may be installed over opening studs, joists, concrete, or concrete block walls with furring strips. Over framing members, use lengths of 1 by 3's for furring strips. Set them 12 inches on center for most tile installations.

The ends of furring strips have to be supported by open framing as shown in this installation. Otherwise, they will wobble and cause damage to the covering—especially ceiling tile. Common or finishing nails may be used to fasten the furring to the framing.

Level or plumb furring strips. You can use cedar shingles for shims to make the strips level. The furring strips have to be uniform throughout the installation; if they are not, the ceiling or wall covering will be wavy. Also, it is difficult to fasten the covering to the strips if the strips are not even.

Standard ceiling tile applications utilize staples to hold the tile to the furring strips. Some ceilings have a metal grid system which is suspended from the framing. Here, the tiles are laid into the metal frames; gravity holds them in position. You can buy double tile that is 12 by 18 inches, or single tile that is 12 by 12 inches. There is a wide selection of tile patterns available in plain and acoustical tile.

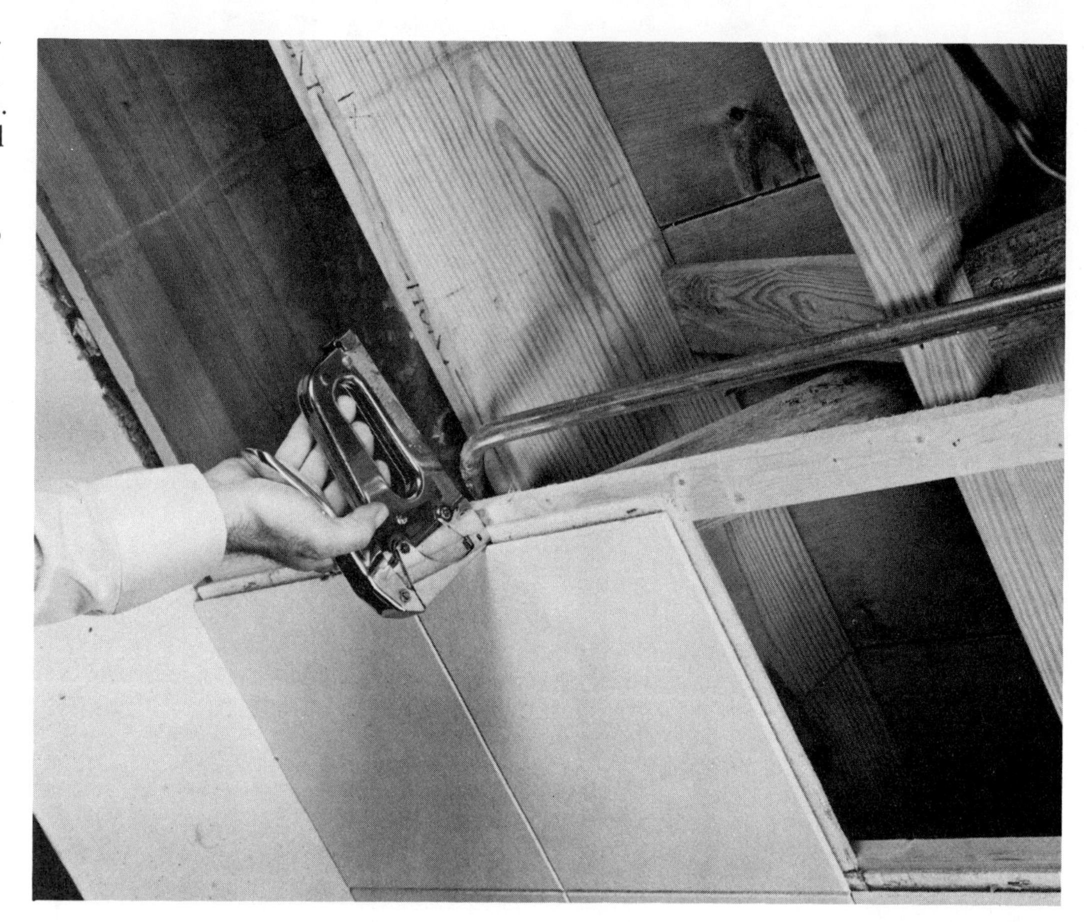

Openings for heat/cooling ducts, lighting, and vents are cut with a keyhole saw. The trim piece on the fixture usually covers the rough edges left by the saw cuts. First, you need to square the tile over the opening; punch or drill a small hole in it and insert the keyhole saw in the opening. Cut around the inside of the duct opening or switch junction box.

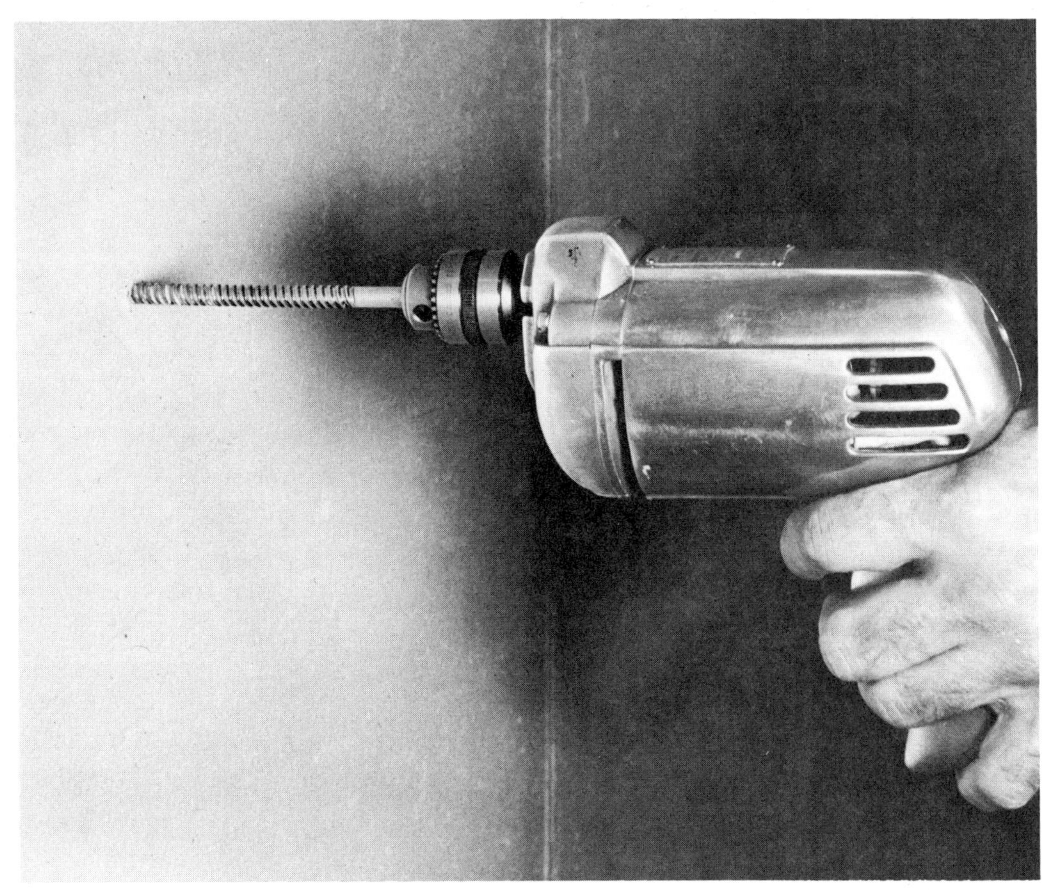

Furring concrete, concrete block walls, or other rigid coverings for paneling require lead or plastic anchors. First drill holes for the anchors. The anchors should be spaced about every 3 feet or so along a line that is snapped with a chalk line. Size of the drill depends on the size of the anchor. The anchor should fit snugly in the hole.

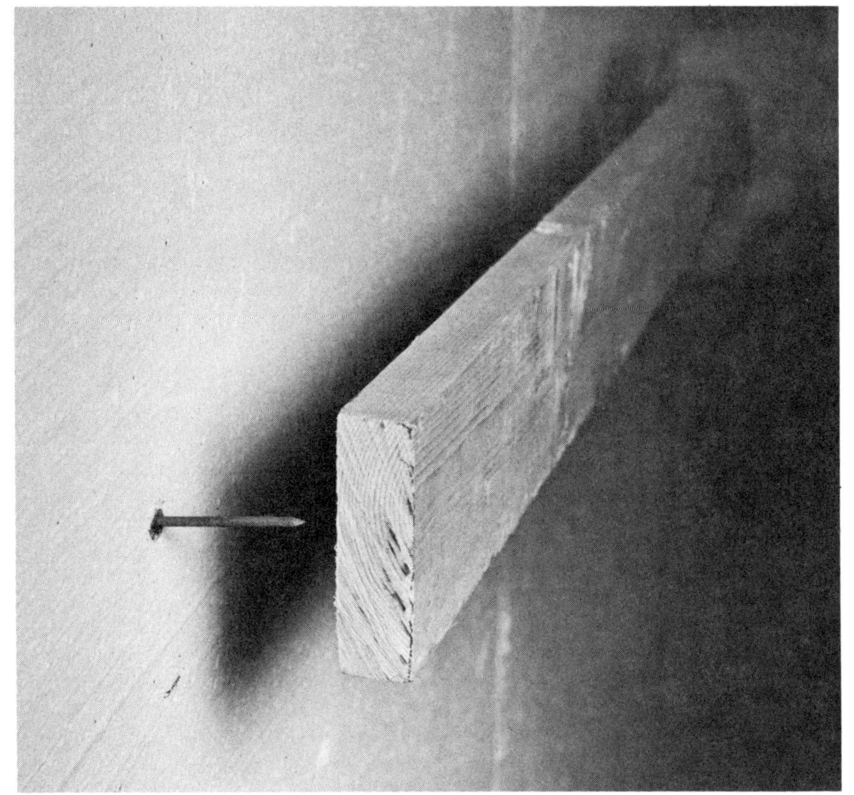

Stick a nail in the anchors. Then level the furring strip over the nails and tap the furring lightly with a hammer. The nails will punch tiny points into the furring strip; these points indicate where holes should be bored for the screws that hold the furring to the wall. Space the furring strips about 3 feet apart or according to paneling manufacturer's recommendations. You must have proper bearing surface. If need be, you can shim the furring strips plumb with cedar shingles. The strips also must be level (or vertical).

Lead anchors for the screws that hold the furring strips are tapped into the holes in the wall. Plastic and fiber anchors are also made for this job, but, unless they are fairly closely spaced, they may not hold the weight of the furring. On smooth walls, paneling may be fastened directly to the wall with adhesive. Adhesive also may be used to install ceiling tile if the ceiling surface is covered and smooth.

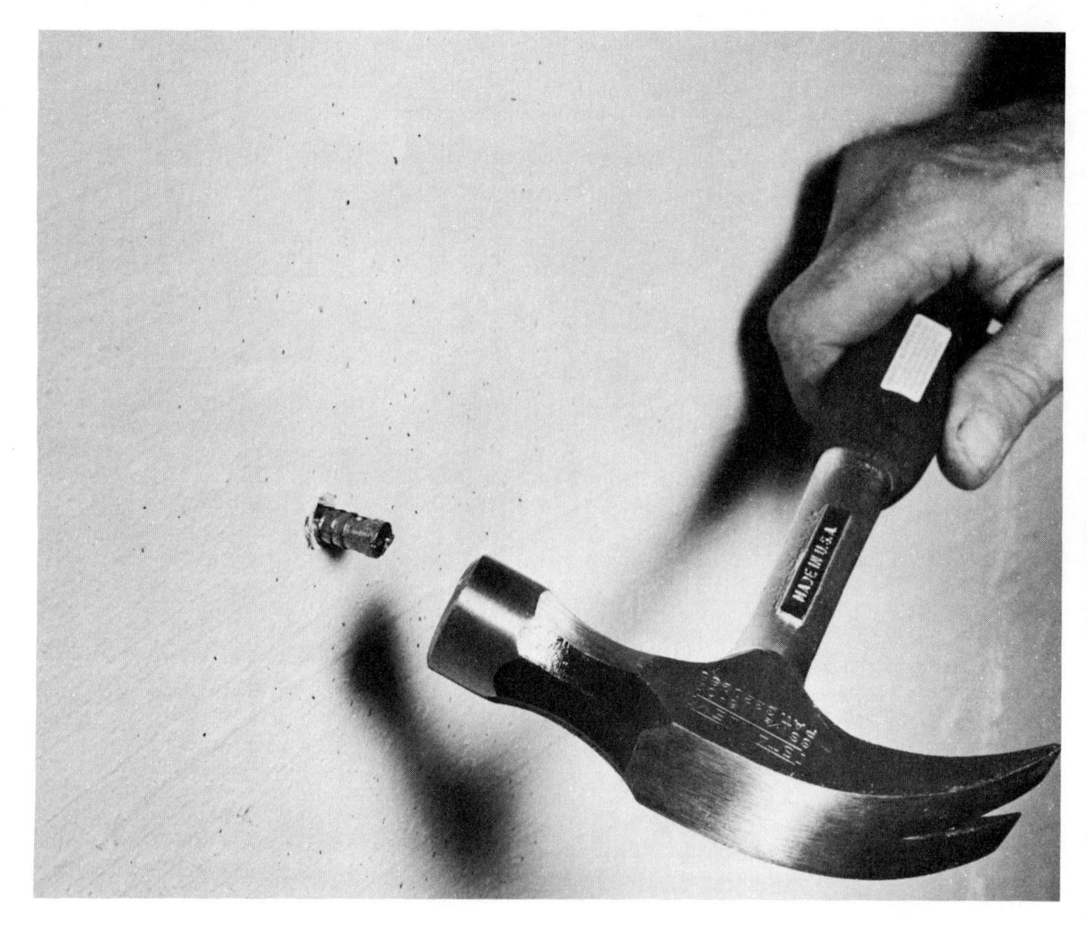

Index

A

Abrasives, 65
Adhesives, 51-62
Adjustable wrench, 46
Allen wrench, 47
Auger bit, 37
Awls, 33, 35, 41

B

Backsaw, 21
Ball peen hammer, 13
Bar clamps, 49
Bearing wall headers, 77
Bevel gauge, 44
Bits, 40
  Auger, 37
  Expansive, 39
  Fly cutter, 40
Block plane, 28
Bolts, 54-56
  Molly, 54
  Toggle, 55
Box wrench, 47
Braces, 37
Bradawl, 33, 35, 41
Building materials, 63
Butt joint, 58

C

Caliper rule, 44
C-clamps, 48
Ceilings, 7, 90, 91, 92
Chalk line, 43
Chisels, 22-25
Clamps, 48
  Bar, 49
  C-clamps, 48
  Pipe, 49
  Rope, 50
  Spring, 48
  Strap, 49
  Web, 49
Claw hammer, 11
Closet framing, 78
Combination square, 42
Compass, 44
Concrete forms, 71
Coping saw, 19
Corner framing, 77
Countersinking, 14, 39
Crosscut saw, 16

D

Dado joint, 61
Dividers, 44
Drills, 40, 41
  Electric, 38
  Masonry, 41
  Star, 41

E

Edging, 69
Electric drill, 38
Electric saw, 17
Expansive bits, 39

F

Fasteners, 51-62
  Adhesives, 57, 58
  Bolts, 54
  Nails, 51, 52
  Screws, 52
Finishing, 70
Fly cutter bit, 40
Foundations
  Concrete block, 71
  Reinforced concrete, 72
Framing, 70
Framing square, 42
Furring strips, 90

G

Glue block, 48
Gypsum wallboard, 64, 87

H

Hacksaw, 19
Half lap joint, 61
Hammers, 10-15
  Ball peen, 13
  Curved claw, 11
  Masonry, 10
  Ripping claw, 11
  Rubber, 10
  Tack, 14
Handsaws, 16
Hand screws, 50
Hand tools, 9
Hardboard, 64
  Standard, 64
  Tempered, 64
Hardwoods, 63
Heating ducts, 75
Hexagonal wrenches, 47
Hinges, mortising, 23, 24

I

Insulation
  Board sheathing, 75
  Board strips, 74
  Styrofoam, 73

J

Jack plane, 26
Joints, 58, 60, 61, 62
Joists, 80

K

Keyhole saw, 21

L

Lag screws, 56
Leveling equipment
  Carpenter's level, 43
  Chalk line, 43
  Plumb bob, 43
Locking hooks, 8
Lumber, 63
  Softwood, 63
  Hardwood, 63

M

Marking gauge, 45
Masonry anchors, 56
Masonry drill, 40, 41
Measuring, 42
Measuring tools, 42, 43, 44
Metal hangers, 85
Miter box, 21
Miter joint, 60
Moldings, 64, 69
Mortise cuts, 23
Multibladed plane, 27

N

Nails, 51-52
Nail set, 14, 15
Needlenose pliers, 45
Nut driver, 47

O

Offset screwdriver, 35
Open-end wrench, 47

P

Paneling, 90-94

Partition walls, 74
Phillips screwdriver, 34
Pipe clamps, 49
Plane iron, 30
Planes, 26-31
  Block, 28
  Jack, 26
  Multibladed, 27
  Smoothing, 26
Pliers, 45, 46
Plumb bob, 44
Plumbing/wiring, 79, 85
Plywood, 65
  Hardwood-faced, 63
  Lumber-core, 65
  Particle board core, 66
  Softwood-faced, 63
  Veneer-core, 65
Plywood spacers, 80
Power tools,
  Electric drill, 38
  Electric sander, 31
  Power saw, 17
Push-button wrench, 47

R

Rabbet joint, 62
Rabbet plane, 10, 26
Ratchet brace, 37
Ripsaw, 16
Roof rafters, 83, 84
Rope clamps, 50

S

Safety tips, 65, 67
Sanders, 31
Saws, 16-21
  Backsaw, 21
  Coping, 19
  Crosscut, 16
  Electric, 17
  Hacksaw, 19
  Keyhole, 21
  Ripsaw, 16
Screwdrivers, 32-36
  Blades, 32, 36
  Offset, 35
  Phillips, 34
  Screw-holding outfit, 33
  Spiral ratchet, 36
Screw nut, 29
Screws, 52
Scribers, 45
Sharpening stones, 25
Slip-joint pliers, 45, 50
Smoothing plane, 26
Socket sets, 47
Socket wrench, 47
Soffits, 86
Spackling compound, 88
Spade bit, 40
Spiral ratchet screwdriver, 36
Spring clamps, 48
Squares
  Combination, 42
  Try, 38, 42
Star drill, 41
Strap clamps, 49

T

Tack hammer, 14
Toenailing, 71
Tool board, 7
Try square, 38, 42

V

Vise-grip pliers, 46

W

Wall fasteners, 54, 55
Washers, 52
Web clamps, 49
Windows, 76, 81, 82, 83
Wood-boring bits, 40
Wooden mallet, 10
Wood joinery, 57
Workbenches, 7
Workshops, 5-8
Wrenches, 45, 47
  Adjustable, 45
  Allen, 47
  Box-end, 47
  Open-end, 47
  Socket, 47